IT'S IN YOUR MATTRESS, YOUR CAR, YOUR COMPUTER,
YOUR CLOTHES, YOUR PAPERS AND YOUR BRIEFCASE.

IT HAS TRANSFORMED YOUR LIFE AND YOUR WORLD.
AND YOU MAY NOT REALIZE WHAT'S EVEN THERE.

STUFF

The way our world is, how it got there and where it's going, is a direct result of the stuff we make other stuff out of: the metals, composites, ceramics, plastics and semiconductors found in every man-made thing around us. The invention of mass-produceable steel fueled the industrial revolution and two centuries of war. Printer John Wesley Hyatt's trial-and-error creation of celluloid kicked off a plastics revolution. Soon, scientific engineers will give us building materials that "heal" themselves and luminous polymers that could be used to create a television screen that rolls up like a poster.

From antique china to airplanes, transistor radios and supercomputers—from the Stone Age to the Electronic Age and far beyond—science writer Ivan Amato takes us on a journey through a breathtaking universe of discovery and challenge; revealing the secrets, the astounding history, the glorious future and possibilities of *Stuff*.

"No one has done a better job in conveying the importance
and excitement of the science of materials...
STUFF is a smart and sweeping guide
to how we have arranged the matter of our universe,
from our earliest toolmaking hominid ancestors
to the wizards who create clean-room tapestries atom by atom.
We learn the fascinating stories of the people who created
the stuff we take for granted, and we get a look inside
the labs of today's molecular engineers."
—David Voss, Senior Editor, *Science Magazine*

"With great enthusiasm and clarity,
Amato scans the process of finding out what stuff does
and then manufacturing something out of it...
[He] provides an evocative explanation
of the engineers' excitement."
—*Booklist*

STUFF

The Materials the World Is Made Of

IVAN AMATO

AN AVON BOOK

To Mom and Dad

AVON BOOKS, INC
1350 Avenue of the Amencas
New York, New York 10019

Front cover photograph by Arthur Holeman/INTERNATIONAL STOCK and Dee Breger
Insıde back cover author photograph by Mary Amato
Publıshed by arrangement wıth BasıcBooks, a dıvısıon of HarperCollıns Publıshers, Inc
Vısıt our websıte at **http://www.AvonBooks.com/Bard**
ISBN 0-380-73153-3

The BasıcBooks edıtıon contaıns the followıng Lıbrary of Congress Catalogıng ın Publıcatıon Data

Amato, Ivan
Stuff the matenals the world ıs made of / by Ivan Amato
p cm
Includes ındex
1 Matenals—Popular works 2 Hıgh technology—Popular works I Tıtle
TA403 2 A48 1997 96-37051
620 1'1—dc21 CIP

Fırst Bard Pnntıng September 1998

Pnnted ın the U S A

OPM 10 9 8 7 6 5 4 3 2 1

CONTENTS

ACKNOWLEDGMENTS

The path to this book is populated by people to whom I owe an enormous debt of gratitude. I thank my teachers throughout life for encouraging my curiosity and for helping me learn how to learn. Special thanks to my fourth-grade teacher, Mr. Wallace, wherever you and your motorcycle are. Special thanks to my ninth-grade biology teacher, Mrs. Hershey, whose perpetual enthusiasm probably enriched the lives of more people than she is aware of. Thanks to Dave Roberts, a longtime friend whom I first got to know when he was my college advisor and organic chemistry teacher at Rutgers College; he convinced me at the impressionable age of eighteen that I actually might have a working head on my shoulders. Thanks also to professor Renee Weber for engraving on my mind that the stuff of the world is as good an altar as one in a cathedral. Finally, in the teacher category I must send heaps of thanks to my one and only journalism professor, Holly Stocking, who helped me believe that I could actually become a science writer.

In the writing world I thank Joel Greenberg and Laurie

Jackson-Vaughan, both of them former editors of *Science News*, for taking me on as an intern at the much-loved publication in the summer of 1986. That experience became the ticket into the science writing world I so badly wanted to enter. I thank Curt Suplee of the *Washington Post* for actually printing my articles in one of the nation's premier newspapers, an honor that I at one time thought was never within my reach. Thanks also to Nancy Enright of the American Chemical Society for taking me on as green as I was; to Pat Young, a former editor of *Science News* who saw it fit to make materials science a standard topic of coverage and for making me the reporter to cover it; to Tim Appenzeller for his demand for clarity and a relentless ability to see possibilities in stories that many editors would have turned down; and to Mat Heyman, head of the Business and Public Affairs Division at the National Institute of Standards and Technology, for his heroic tolerance of my short attention span.

Since I cannot thank everyone in the science world who I would like to thank, I will name several people who can serve as representatives: Stephen Carr of Northwestern University for a fateful conversation early in my writing career that helped me see how vast and great the field of materials science and engineering is; the late Cyril Stanley Smith, whose poetic and philosophic approach to the world of materials has changed me forever; Federico Capasso, Greg Olson, Mehmet Sarikaya, Paul Calvert, John Angus, Klaus Zwilsky, and Craig Rogers, who have helped me in many ways to bring this book from an idea to a thing that you can throw across the room. I thank the many hundreds of other scientists and engineers without whose collective effort I would never have been able to see the material world in all of its wonder.

I thank Robert Friedel, historian of technology at the University of Maryland, for the skill and diplomacy with which he read early drafts of this book, and for the way his mind works. May my future work emulate his.

In the book world I thank Ivars Peterson, the tireless

reporter at *Science News* who proved to me that writing a book is indeed possible. I thank David Lindley, a book author himself as well as a sometime colleague at *Science* magazine, for hooking me up with Susan Rabiner at Basic Books. And I thank Susan for ushering this book slowly but surely from the raw ore that she encountered initially into a handsome edition that I can rest assured will sit upon my parents' bookshelf.

Which brings me to my family, whom I thank most of all. To my father, Sol, and my mother, Sylvia, thank you for literally making me and for helping me to see that the world *and* our ability to know it at all are infinitely awesome phenomena. There is no greater gift that parents can give to their children.

And, finally, I thank my wife, Mary, and my two young boys, Max and Simon. I thank Mary for sharing every single up and down that goes with writing a book. I thank her for her infinite supply of encouragement, goodness, and love. To Max and his little brother Simon, I thank you both for coming into this world and for reminding me every day what really is important.

INTRODUCTION

STUFF, *stuff*, *Everywhere* STUFF

Stop reading this book. Instead, hold it in your hand. Feel its weight. Run your finger over a page and feel its smooth texture, its dryness. Sense the page's compliance to the pressure of your finger. See how its whiteness is punctuated with black symbols printed in ink in a few dozen rows. If you sniff the page, you even may smell some aromatic chemicals still fleeing from the ink. Rub the page between your thumb and forefinger and hear that familiar papery noise. Twist a corner and see how the paper crinkles. Rip the page and see how the ragged edges hint of the tiny wood fibers that became entangled at a mill to become the paper.

Now examine the book's fabric spine to which the pages are attached with an adhesive. The stack of pages, in turn, attaches to the back and front covers of the hardcover edition by way of sheets of heavier-gauge paper that flank the book's first and last pages. At the bindery a machine pasted these flanking sheets to the rectangular pieces of stiff, heavy paper-board that the front and back covers are made of. These heavy covers, themselves enveloped with a sheet of paper or polymer-impregnated cloth, are attached to each other by a flexible paper or fabric bridge that ends up directly over the spine so that readers can open the book. The softcover edition is a simpler affair of paper, ink, and glue.

Soft or hard, this book is made through and through of stuff.

And so is every thing else. Human beings extract about 15 billion tons of raw material—that's 30 trillion pounds—from the earth each year, and from that they make every kind of stuff that you can find in every kind of thing. Mined ore becomes metal becomes wire becomes part of a motor becomes a cooling fan in a computer. Harvested wood becomes lumber becomes a home. Drilled petroleum becomes chemical feedstock becomes synthetic rubber becomes automobile tires. Natural gas becomes polyethylene becomes milk jugs and oversize, multicolored yard toys. Mined silica sand become silicon crystal becomes the base of microelectronic chips. Each kind of stuff is a link to enormous industrial trains whose workers process the world's raw materials into usable forms that constitute the items of our constructed landscape.

Each kind of stuff also is a palimpsest of innovations in the use of materials, some going back to prehistoric times. The wood-pulp paper from which books are made today comes from a pedigree of cotton and linen rags, animal-skin parchment, Nile-reed papyrus, and Sumerian clay tablets. The ink, a black pigment made from the ground ash of some carbon-bearing fuel and then suspended in a rapidly evaporating solvent, has its roots in crushed ore and charcoal mixed with spit or animal grease for use on cave walls and faces. The materials in every book tell a tale that rivals the one conveyed in its words.

It is the same for every other material thing that you encounter. Train your attention on the stuff of things rather than on their function. What you see is a rich medley of materials: the liquid crystal display of your laptop computer; the gritty concrete sidewalk on which you are strolling; the nylon of your raincoat's zipper; the carbon-fiber-reinforced epoxy polymer of your tennis racket; the Kevlar polymer in your police force's bullet-proof vests; the oak of your dresser; the diamond in your engagement ring; the nickel-based superalloy in the turbine blades in the engine of an airliner you are flying in; the warm, supple skin of your newborn; the cool, transparent glass of your office window; the combination of slick, high-density polyethylene and stainless steel that make up the artificial hip which a surgeon may have implanted into you; the cotton of the shirt you are wearing; the aluminum of the can you just drank from.

In a single day the thousands of man-made materials that you encounter, engage, manipulate, and use display a diversity every bit as wondrous as that found in living organisms, which are composed of the most miraculous of all the world's materials—skin, bone, tendon, muscle, nail, hair, and scads of other biological tissues—all of them honed by evolutionary engineering into a beautiful marriage of form and function.

That books and buildings and the things in the world are supposed to be made of materials suited for their functions is so obvious that it almost goes without saying. But things that go without saying long enough are readily forgotten. That is why the materials that make up the world are most often not on people's minds.

When the stuff of the world does comes to the fore, it is often a consequence of a material failing to serve its intended role. This happened in 1988, when the aluminum-alloy skin of an airliner flying over the Pacific Ocean peeled open and sheared off like the top of a sardine can, sending a flight attendant to her death. Articles about metal fatigue appeared in the *New York Times* and all over the media. Aluminum and material failure enjoyed their fifteen minutes of fame. To the unseen scientists, engineers, and workers who actually transform the raw stuff of the world into materials like aerospace alloys, the way materi-

can go bad is never far from their minds. They—and their bosses, lawyers and insurance companies—know how lives, economies, and national security depend upon understanding what makes materials the right stuff or the wrong stuff.

But materials are far too central to our lives and too fascinating to remain out of mind until they become sensational headlines. Paying attention to the materials in your life spawns an endless stream of fascinating questions (and even more fascinating answers). Why is a skyscraper's frame made of steel, not copper or polyethylene? What is so special about silicon crystal that it became the foundation of a society-changing electronics revolution? What kind of materials does it take to build a plane that can take off like a jet, accelerate at 20,000 miles per hour, zoom into outer space, and then land two hours later in an airport on the other side of the planet?

Simply noticing the diversity of stuff in the material world can evoke an openmouthed sense of wonder akin to visiting a zoo filled with animals that you had never really seen up close. That sense of wonder might even grow into awe when you consider this: the constructed world brims with metals, ceramics, plastics, fabrics, glass, and thousands of other specific materials that are found nowhere in the wilderness.

And that sense of awe naturally leads to a question like this one, which is at the heart of this book: How in the world did humanity learn to transform the raw stuff of the wilderness into the contemporary zoo of materials? Nylon doesn't grow on trees. It starts out in a barrel of tarry, smelly coal tar or petroleum. Someone—in this case a chronically depressed and brilliant chemist who ultimately committed suicide—figured out ho[illegible] pull off the transformation. Someone had to figure out [illegible]onvert a handful of sand into an optical fiber that can [illegible] around for miles and miles the way a copper wire [illegible]ity.

[illegible] wrote that you cannot know something fully [illegible] it from its beginnings. This applies to the human use of materials. So in this book, I will start from the beginning—about 2.5 million years ago. That is when our hominid progenitors in the Great Rift Valley of Africa likely had the original insight that materials as they are found in the wilderness can be transformed into more useful and empowering forms.

The original noisy act of materials engineering—striking a stone against another to produce a sharp, cutting edge where before there had been a dull surface—is no different in its core from what modern engineers in Silicon Valley do when they implant boron and phosphorous atoms (stepping stones for electrical charges) into pristine wafers of crystalline silicon to create the semiconductor of the microelectronics revolution. The goal of both procedures is the same: to alter materials so that they can perform in new technological ways.

Despite the apparent crudeness of chipping stones to make tools, the kinship between the original stone flakers and the most modern materials engineers goes atomically deep. Silicon atoms, when combined with oxygen atoms to form specific minerals, is the basis of one of the most coveted materials of the first stone toolmakers—flint. Because of the arrangement of internal mineral grains in flint, the stone can cleave conveniently in a shell-like pattern that leaves sharp edges. Moreover, its edges are hard and tough enough to remain sharp even after cutting and scraping into materials like tendon and bone. Silicon has been a celebrity element from the beginning.

From the Great Rift Valley, the story of materials progresses through some 125,000 hominid and human generations (based on a twenty-year generation) to the present day. During that vast stretch of time, the first hominid toolmakers evolved into us, *Homo sapiens*. Punctuating those thousands of millennia were a few major turning points, each leading to entirely new powers of material transformations that have changed the course of individual lives and of global history. (See Figure 1.)

The use of fire to change the stuff of the world was one of

them. Fire's heat has the capacity to reach down to the molecular identity of a material and then to jumble, reorient, and rearrange it to yield almost magical transformations. Wood becomes smoke, heat, invisible gases, and ash. Dull, crumbly ore becomes lustrous metal. Wet, malleable clay becomes hard pottery. Sand become glass. It may all have been accidental chemistry in the hands of the first fire users, but the consequences were and are world-changing.

Another major turning point was the development of a more deliberate type of chemistry and its far more extensive ways of reaching into and transforming the anatomy of stuff. For centuries alchemists, chemists, and sundry tinkerers pulverized, mixed, boiled, distilled, roasted, extracted, electrolyzed, and otherwise fiddled with whatever substance they could find in the mineral, animal, and vegetable domains. Out of this obsessive quest came a myriad of previously unseen substances—solid, liquid, and gaseous—some of which harbored uses by themselves or as participants in new chemical reactions.

Along the way, chemists also learned about the limitations of their business. Despite the two-thousand-year alchemical quest to transform "base" materials into gold, no one ever succeeded in transmuting one element into another. At least on earth, under normal conditions (that is, without enormous

FIGURE 1

The Stuff of History. To follow each of the heavy lines across this diagram is to trace a word-byte history and projection of four major classes of materials—*metals, polymers-elastomers, composites* (whose properties are derived from two or more materials—such as a polymer with glass fibers as reinforcement—bound into a single overall structure), and *ceramics-glasses*. The relative importance to humanity of each material class at each historical period is suggested on the diagram by the proportion of the diagram's vertical axis corresponding to each material class. So in 1960, for example, composites were least important while metals were head and shoulders most important. In 2020, however, the four classes of materials are projected to have roughly equal importance.

MICHAEL F ASHBY

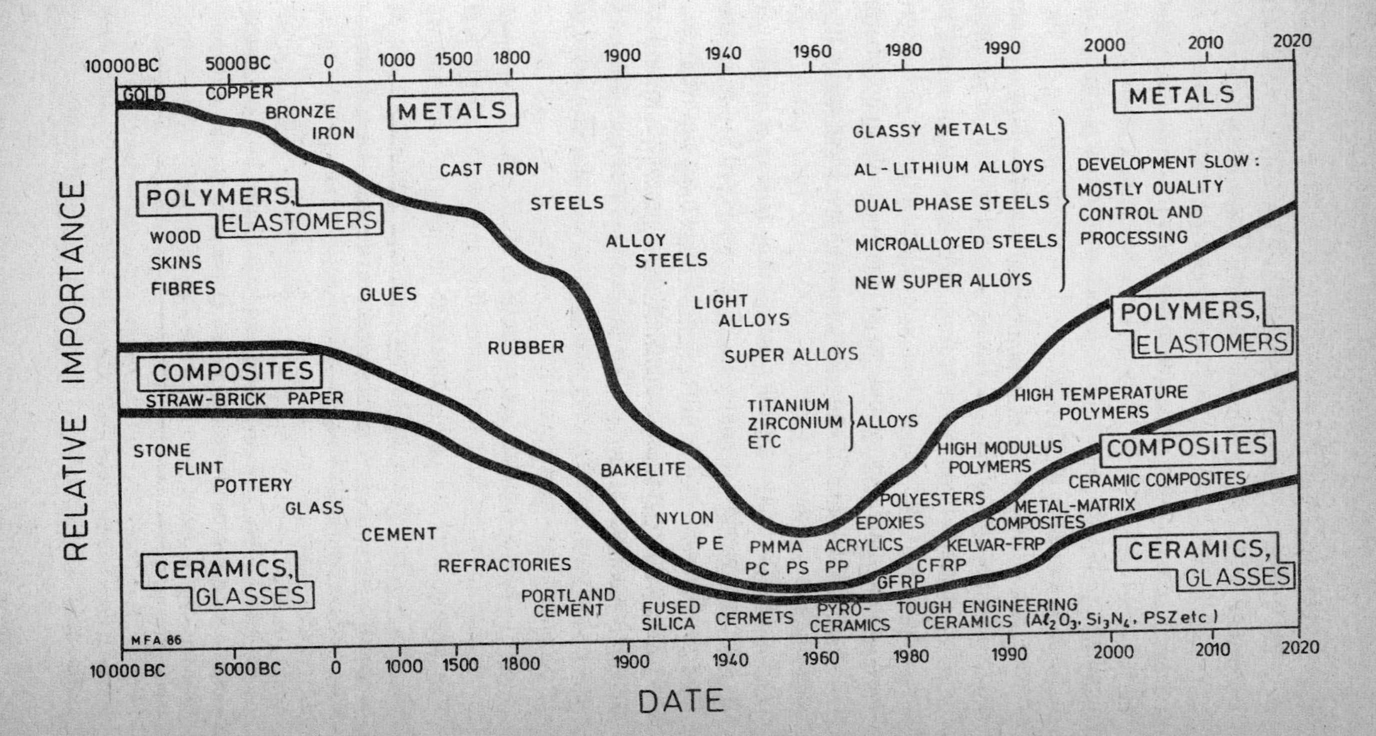
10000 BC
5000 BC
0
1000
1500
1800
1900
1940
1960
1980
1990
2000
2010
2020
GOLD
COPPER
BRONZE
IRON
METALS
CAST IRON
STEELS
ALLOY STEELS
LIGHT ALLOYS
SUPER ALLOYS
TITANIUM ZIRCONIUM ETC
ALLOYS
GLASSY METALS
AL - LITHIUM ALLOYS
DUAL PHASE STEELS
MICROALLOYED STEELS
NEW SUPER ALLOYS
DEVELOPMENT SLOW: MOSTLY QUALITY CONTROL AND PROCESSING
METALS
RELATIVE IMPORTANCE
POLYMERS, ELASTOMERS
WOOD
SKINS
FIBRES
GLUES
RUBBER
BAKELITE
NYLON
P E
PMMA
PC
PS
ACRYLICS
PP
EPOXIES
POLYESTERS
HIGH MODULUS POLYMERS
HIGH TEMPERATURE POLYMERS
POLYMERS, ELASTOMERS
COMPOSITES
STRAW-BRICK
PAPER
GFRP
CFRP
KELVAR-FRP
METAL-MATRIX COMPOSITES
CERAMIC COMPOSITES
COMPOSITES
STONE
FLINT
POTTERY
GLASS
CEMENT
REFRACTORIES
CERAMICS, GLASSES
PORTLAND CEMENT
FUSED SILICA
CERMETS
PYRO-CERAMICS
TOUGH ENGINEERING CERAMICS (Al_2O_3, Si_3N_4, PSZ etc)
CERAMICS, GLASSES
MFA 86
DATE

accelerators), scientists realized they would never become alchemists—they would never change lead into gold or oxygen into carbon or any one element into any other element.

This limitation, it turns out, was always more a state of mind than of reality. The hard-won insight that the vast menagerie of materials in the world is a result of a small pantry of elements, that the personality of each material is the result of an inner hierarchy of chemical and physical structures, has became the principle of a new alchemy practiced by people known as materials scientists. Rather than gold from lead, their trophies are numerous, ranging from new lightweight alloys to build a next-generation fleet of superefficient jetliners to harder-than-diamond materials for industrial machine tools to polymeric drug-soaked medical implants that can release their cargo into diseased brain tissue.

Yet another major turning point in the human control over the material world—perhaps the last one possible—is in the offing. In the past half century a field known as materials science and engineering has emerged as a powerful hybrid of many other technical fields. The practitioners of this field are coming to a point where they are gaining the ultimate level of control over the material world.

No longer contented to manipulate "stuff," they have tools for seeing, moving, and understanding individual atoms and the ever larger atomic collections that become the materials we use. The emerging ability to micromanage materials even at the atomic level is giving researchers unprecedented access to the mostly untapped material wonders of the periodic table of the chemical elements. Even compared to today's most sophisticated materials like the semiconductor in a computer chip or the gallium arsenide in a CD player's laser, materials scientists say we still ain't seen nothing compared to the materials that will come.

Materials scientists now recognize that within their new powers lie new solutions to many of the globe's most vexing problems—among them pollution, energy supply, housing, transportation, communications, and poverty: new superconductors that can dramatically increase the efficiency with which electricity is generated, distributed, and used could demote petroleum as a factor in geopolitical dynamics; raw materials for making these superconductors—say, ores containing unfamiliar elements like yttrium and bismuth—could become strategic materials; new metallic alloys such as titanium-aluminides and new composite materials—perhaps ones made of tough ceramic materials laced with superstrong diamond fibers that overcome the ceramic's endemic vulnerability to catastrophic fracture—will be the stuff of new aircraft that can fly at twenty-five times the speed of sound.

"Smart materials" that can respond to external conditions by changing their color, shape, stiffness, or permeability to air or liquids could become the stuff of future cities whose buildings are more comfortable and better able to field sudden violent challenges from earthquakes or terrorist bombs. Smart materials also could lead to cities whose infrastructure can sense—and even automatically compensate for—the wounds of corrosion, metal fatigue, age, and the other slings and arrows of urban decay.

On the down side is that today's wonder materials could become tomorrow's environmental villains. It has happened before. When chlorofluorocarbons (CFCs) were first invented in the 1930s, their combination of inertness, nonflammability, and thermophysical properties fated them to become materials of choice for technologies such as refrigeration (which previously relied upon extremely noxious materials including ammonia and sulfur dioxide), aerosol propellants, and blowing agents for making lightweight materials such as styrofoam.

The inventor of CFCs, Thomas Midgely, used to breathe CFCs into his lungs and then extinguish a candle with a CFC-rich exhalation to impress audiences with the material's benign nature. Over the course of decades of ever higher volume use, however, CFC molecules had been meandering into the stratosphere where, it turns out, they trigger reactions that destroy ozone molecules. Ozone molecules absorb much of the sun's life-unfriendly ultraviolet radiation. So what were once new wonder materials have become one of the century's most prominent chemovillains.

Today's materials researchers are far more aware of the Promethean catch that new materials can harbor, having learned from experience with CFCs, lead-based paint pigments, asbestos insulation, and many other materials. The challenge for today's researchers is to use their ever more sophisticated knowledge base to more accurately predict the balance of potential benefits and costs new materials will bring with them.

In the words of Cyril Stanley Smith, the late materials historian and Manhattan project metallurgist: "Materials themselves have interacted with mainstream history, for they are the stuff on which virtually all human activities are based."[1] As materials researchers discover and create ever more capable materials, human aspiration and achievement will soar into new places.

PART I

Trial and Error

Imagine that every material thing in your life suddenly disappears. Your car. Your home. Your clothes. Your jewelry. Everything, every thing, is gone: the steel, concrete, glass, plastic, and thousands of other materials that made up your modern habitat are nowhere to be found in ready-to-use form. Naked in the wilderness, you are going to have to figure out how to transform the raw materials of the world into forms suitable for making things. Your face contorts with the anguish of profound puzzlement.

Your mind focuses on your first goal: covering your nakedness. The wilderness becomes a primordial hardware store with resources in the rough. The first material you consider for making clothes is some birch bark, thin and peeling from its tree. Stitch a few patches of this stuff together, and just maybe . . . But when you get hold of the bark (whose oil you might later discover to be helpful in starting fires), you find that the bark is too stiff. Besides, you have no idea what you could use to stitch pieces together. Won't work. Next, you catch sight of some large leaves, green and supple. But you find that they rip too easily. Even if you finessed them into a loin covering, they would rip apart with your first step. Two trials, two failures.

As you scan the landscape for still another suitable material, a deer streaks across your path. Its beautiful hide seems to scream, "Wear me." Notwithstanding the animal rights movement of your former world, you want nothing more than to hunt down that deer and make clothes out of it. You have no gun, no bow, no arrows. Stones come to mind, and sharpened sticks. To find them you scour the forest floor for stones that have the right feel. They must be light enough to hold and throw but heavy enough to stun a deer. You try hundreds of stones and end up with a dozen that seem good for hunting. As for sticks, most are rotten, and many break too easily when you bend them. Some are strong enough, but they are too green to sharpen.

Next to a rotting log, however, lies something far more promising. It is an antler shed by a young buck. Its tips are sharp, hard, and strong; all at once you intuit the possibilities. If you could find a way to haft one of the straighter prongs of the antler onto a wooden shaft, you would have yourself a nice spear. The thought pleases you, but your confidence dips again when you acknowledge that you must still cut the antler into suitable pieces and also locate a good shaft. A sharp-edged stone might work for sawing through the antler. . . .

And so it would go on. For each task of survival at hand, you would have to find the kinds of materials, technologies, and techniques suitable for the job. It would be a painstaking exercise of trial-and-error, perhaps punctuated by serendipitous godsends like finding a perfectly straight antler tip or a nature-made stone blade. By enduring enough trial and failure, you would learn crucial lessons about the materials in your environment. Gradually, necessity transforms you into a craftsman.

Trial and error and accidental discovery: these are the procedures by which for the last 2.5 million years, humanity and its hominid progenitors slowly have acquired the skills needed to transform raw materials into stuff for constructing things—for the first makers of stone tools; for early metallurgists who learned to combine copper and tin ores to make more workable and harder bronze alloys; for the Samurai sword makers, who learned and then ritualized the transformation of iron into the superior steel-edged weapons that made the Samurai culture the stuff of legends; for the glassmakers on the island of Murano, next to Venice, who happened onto just the right mix of ingredients and just the right furnace designs to make glassy materials with an elegance and unprecedented transparency for which the island became famous five hundred years ago.

Alchemists also practiced trial and error—with perhaps a greater emphasis on error—in their obsessive search for the philosophers' stone, the esoteric preparation they believed would transform lead into gold. In one of the most notable instances of trial and error applied to an impossible goal, they putrefied, boiled, distilled, solidified, and sublimated all kinds of mineral and organic ingredients in every possible sequence. How could these eccentric tinkerers, Isaac Newton among them, have known that nothing short of nuclear physics in huge accelerators and reactors could pull off such a fundamental transformation?

To this day in a scientific age, trial and error remains a cornerstone in the quest to discover and invent ever more capable materials for making things faster, stronger, more durable, and otherwise better than before.

ONE

EARTH, *Air*, FIRE, AND WATER

One day about 2.5 million years ago, a protohuman living in eastern Africa's Rift Valley made a momentous discovery. No one knows the details, but it could have gone something like this:[1]

During a search for rocks suitably sharp-edged for scraping meat from bones, this hominid—either an *Australopithecus* or *Homo habilus* whom we shall call George—picked up a hunk of dark brown flint by a chalky cliff face. Finding it too dull, he threw it down, an exercise he and his kind had repeated a million times before. This time, however, it was different.

The loud report of the stone hitting another stone drew George's attention. Right there at the point of impact lay just what he had been looking for—a sharp-edged flake the size of his palm. As George stared at this unbidden gift, his brows furrowed. A novel shiver of insight swept over his nearly human brain. Seizing another piece of flint, he once again struck rock against rock, and more sharp-edged flakes laid themselves out in front of his discovery-widened eyes.

In one stroke George unwittingly launched the Paleolithic Age (the Old Stone Age) and invented materials engineering. Materials engineers are the people who figure out how to make the estimated fifty thousand materials that compose the modern industrial landscape, including the silicon crystals of computer chips, the ultrapure glass strands of optical fibers, the nickel-based superalloys of jet engines, and the synthetic rubber of car tires—none of which are found lying around in the wilderness. They are made the way George made his first artificial stone tool: by transforming the raw materials of the world into new and more useful forms.

Before George's discovery technology had always been the art of using things pretty much as they were found. A stick from a flat-topped acacia tree was good for getting at ants underneath the savanna's hard earth. A heavy, egg-shaped lava stone was good for hammering open wildebeest bones to get at the marrow-rich interiors. Left intact, the same bones were good for beating the daylights out of the living flesh of a rival protohuman. The Old Stone Age, which lasted until about fifteen thousand years ago, certainly also was the age of wood, bone, hide, leaf, shell, hoof, horn, tendon, tooth, tortoise shell, claw, vine, bark, sap, and of any other easily accessible material that proved useful in its raw form.

George, our Paleolithic Edison, realized that hominids like him did not have to be satisfied with the world as they found it. Bashing one rock against another can get you a sharp cutting edge where there wasn't one before. And with a ready supply of sharp stone tools, George could scavenge more meat from recently killed prey in the little time he had before other scavengers like hyenas arrived with their own lust for the same meat. Stone flaking meant larger and tougher-skinned animals could be butchered, which made more of the world habitable. Early stone technology helped open doors out of Africa to Europe, Asia, China, and eventually to North and South America. George's discovery was worthy of a special Nobel

Prize given every few hundred thousand years: a lifetime of premium flint to the winner.

The technology of stone flaking, dubbed the first technology by some paleoanthropologists, probably spread quickly throughout George's own clan, and then more slowly to other clans. Toolmakers found that certain ways of hitting stones together produced more useful shapes and sharper edges; they found that materials like flint and obsidian (volcanic glass) were better suited to making tools than crumbly pumice or overly hard granite.

The clacking of stone against stone was the sound of high technology for so long that George's descendants had enough time to evolve into modern humans. Ice ages came and went; deserts greened into forests and dried back into deserts; entire animal and plant species emerged and disappeared; but the stone tools themselves didn't change all that much.

During all of this time, there were, of course, other Paleolithic Edisons. Hominid stone workers at various times and places came up with variations on the original theme. But that theme hardly changed in 2.5 million years: start with a suitable stone blank and remove bits and pieces until you end up with a scraper, a hammer, an ax, a blade, or some other tool. For some 124,000 Paleolithic generations, innovations in the use of materials were more or less footnotes to George's happy accident.

During this epoch the effect of materials engineering on the look of the landscape remained minuscule. Wherever they went and lived, the toolmakers littered the landscape with their stone hammers, scrapers, choppers, hand axes, piles of flint debris, and the bony remains of their meals. Like materials engineering, litter is a Paleolithic invention. But this litter was small scale. Even today's best spy satellites, the ones that can read a license plate from orbit, would have stood a slim chance of detecting signs of the bipedal creatures who were slowly expanding their dominion by devising ways of engineering the stuff of their world.

Human exertion can fracture a stone along weak internal pathways that developed, like biological traits, during the stone's geological genesis. When Paleolithic hominids did this skillfully, they got nice sharp flakes or beveled edges, all the better than muscle and teeth for butchering and other daily chores. But God-given anatomy alone has never allowed a hominid or human to reach into the microscopic interiors of raw materials and rearrange their chemical bonds (leaving aside the miracle by which digestive systems transform food into living body tissue, biochemical energy, and excrement)—to convert clay into pottery, ore into metal, or sand into glass. That requires an entirely different kind of energy. It takes fire.

The harnessing of fire's transformative power may have begun in a place like the limestone Zhoukoudian cave, west of Beijing. The cave was first occupied by *Homo erectus* clans seven hundred thousand years ago and was still in intermittent use five hundred thousand years later. There, in a cavern almost two football fields long and fifty yards wide, hominids made fires. The charred remains of bones, burned stones, and wood—from about 460,000 years ago—pile twenty feet deep in some places. Ironically, using fire for material transformations apparently did not occur to any of the *Homo erectus* denizens of the Zhoukoudian cave, whose use of flame was apparently confined to cooking and obtaining warmth, light, and protection from other animals. By the time the cave was abandoned, however, fire had become part of the hominids' daily experience.

About twenty-five thousand years ago, the modern humans that evolved from the first fire-using *Homo erectus* definitely had discovered what fire can do to the stuff around them. One of the oldest clay objects known to have been deliberately hardened by fire was found in what now is Vestonice, Czechoslovakia.[2] It is a female figurine—an archaic Venus. Formed first out of wet clay, she was endowed by her sculptor with exaggerated sexual anatomy. She was a variation on a widespread figurine theme, carved in ivory and stone, that had extended back at least a few thousand years. But by placing the soft and malleable clay form into the heat of a fire, the archaic artist became an accidental chemist. He or she caused the mineral particles of the clay to transform chemically and physically and bond into a hard and durable material. Clay became pottery.

Utilitarian need was not a likely impetus for making fired clay figurines like the Venus of Vestonice. It seems that the use of fire to change materials into new forms emerged from a religious or aesthetic sensibility. As the late Cyril Smith saw it, necessity has only recently become the mother of invention. Ever since the Venus of Vestonice, he wrote, "the first appearance of almost all artificial materials and of treatments to adjust their properties or shapes to specific uses appear first in objects of art, or at least in objects of ceremony in which utility resides in their aesthetic overtones."[3]

For millennia before and after the making of the Venus of Vestonice and similar prehistoric figurines, the aesthetic sensibility must have been infecting the entire human population like a benevolent virus. Across Europe and Russia artists were carving bone and stone into human and animal forms. In South Asia they were engraving ostrich shells, which they may have used as containers, with simple geometric patterns. In Australia nomadic hunter-gatherers were painting and engraving rocks. In caves in Namibia, Spain, France, and elsewhere, they laid down images of horses, rhinos, ibexes, bisons, and themselves.

In their quest for color, early painters got their black by mixing spit with charcoal from their fires or with black minerals like manganese dioxide, which was easily accessible. White may have come from kaolin, a chalky clay mineral that much later helped craftsmen in China earn worldwide envy for their fine pottery and porcelain. Colorful pigments came from red ocher (hematite that contains iron) and green, copper-bearing malachite. The same atoms that gave these minerals their color—iron and copper, respectively—later became sources for another kind of fire-born material: metal.

The agricultural revolution that began about ten thousand to fifteen thousand years ago marked the end of the 2.5-million-year-long Old Stone Age. What was once high technology—chipping stones into tools—was no longer sufficient for new tasks like sowing the land.

There still were no cities or vast agricultural regions forming grids like huge sheets of graph paper. There were no major roads, pipelines, or aqueducts, whose more geometric lines and contours bespeak human design rather than natural causation. All such advances were still thousands of years away. The human mark on the world remained intermittent and scant, hardly noticeable amidst vast, untrammeled expanses.

Agricultural life, however, meant that more people could live in a smaller region longer than ever before. In this more settled society people could assume more specialized roles. They could observe one another's skills and innovations far more often. There probably was more opportunity than ever before to communicate, commiserate, and search for solutions to problems.

The variety of materials exploited by the early agriculturists most likely expanded at a pace unseen since George invented technology in the first place. People were beating, scratching, grinding, and heating materials to reveal previously hidden possibilities of the things around them. Flake flint and you get a knife. Grind up malachite and you get green pigment. Bake clay and you get pottery. What next?

With the advent of agriculture, the technological limitations of stone must have become irritatingly evident. Granted, stone was still suitable for making grindstones, mortars, and pestles—the tools for making real food out of grains. It also was still the best stuff for small knives, scrapers, axes, weapons, hammers, and other tools. But stone was not a very good material for the rapidly growing need to store, contain, and protect seeds, harvests, and other resources. It was hopeless for transporting water.

For these tasks early agriculturists did have at least some material options. They could shape, carve, sew, or weave hides, gourds, wood, bark, and reeds into containers and baskets. But like all materials, all of these had pros and cons. Even if the raw materials were available in sufficient quantities, the resulting implements were neither waterproof nor insect-resistant. They could get wet and moldy. They could crush and break easily.

Perhaps one visionary saw it. More likely many individuals or groups in various places at various times must have joined in the proverbial Aha! as they made the connection between clay, fire, and the rocklike pottery utensils and vessels that could emerge from these two constituents. Unlike the earlier figurines, the fired clay objects born of this insight were not merely objects of art and ceremony: they were useful in everyday life.

Pottery could be watertight. Insects could not enter pottery vessels as readily as they could more porous baskets. The stuff could be made as long as there was clay, water, and fire. This was one of those big ideas that helped initiate a new era of human history: the Neolithic, or New Stone, Age.

One route of pottery into routine life might be reflected in ancient pieces of fired clay found in Gambles Cave in Kenya.[4] These pieces have wickerlike imprints on them. One explanation for the imprints is that some clay accidentally got smeared on the outside of an ancient basket that ended up in a fire. Or maybe the basket maker used some mud to strengthen the wicker weave, to waterproof it, or at least to keep seeds from spilling out of crevices. Or perhaps someone did in fact smear the mud on the basket with the intention of making a pottery vessel.

Whatever the details and wherever the place, the discovery that fire could transform clay into useful pottery must have been a happy one, like a huge freshwater lake looming before a desperately thirsty hiker. Clay, a kind of earth, was readily accessible to early farmers, particularly along the wet and silty banks of rivers that made the land of the early agriculturists so fertile.

Suddenly there was a seemingly endless supply of the vessels that were increasingly vital to the agricultural pioneers. Unlike any other material humanity had known, pottery's main ingredient, clay, was easily molded in all three dimensions using nothing but hands and water. Once formed, the soft clay shapes could be heated into hard and useful pottery implements. A new material, pottery, was central to the galloping success of the agricultural societies.

Like a fire in a drought-stricken forest, pottery spread quickly from its various origins. Soon fired clay pots, bowls, vessels, pipes, and other items for cooking, eating, storage, and hauling became so prevalent that their remnants have become the ancient "paper trails" by which historians and anthropologists have re-created who went where and when.

As ever more people began working with clay and fire, moreover, a form of positive feedback kicked in to quicken the pace of innovation with materials. In different places early potters dug differently shaped hearths and used different fuels (wood here, dung there) and different raw materials. The result was a diversity of pottery types: different qualities, styles, colors, strengths, and porosities.

Pottery, useful at it was, also had its flaws. Those same broken shards of pottery that have attracted armies of scholars and museum curators are testimony to pottery's great limitation: knock a pottery bowl and it chips; drop it and it shatters. The cringe on a contemporary face witnessing a teetering plate or glass surely mirrors the pained expression of an early Neolithic potter beholding yet another shattered masterpiece.

Pottery's fragility arises from the way the countless mineral particles that make up clay become a single pot by way of fire. Each clay particle is a minuscule pebble—a tiny grain that may be a single crystal or an interlocking conglomeration of many separate minicrystals. In a hot enough fire, the grains sinter, which means that they bond together without melting. The grains do not, therefore, weld together seamlessly without breaks. Instead, they end up sporadically stitched together via chemical bonds between adjacent grains.

The result can be and often is quite strong amidst compressive forces. Four high-quality Wedgewood teacups can support a tank. But woe to those who put a great deal of tension on pottery vessels or teacups: such tension entails a force that pulls on the chemical stitching holding the grains together. That happens when things fall or are hit or bent. When a plate falls off of a table and hits the floor, some of the energy of the impact pulls on the grains of the plate, instantly opening tiny cracks between grains, which then pry apart more and more grains at an accelerating rate. When cracks reach from one end of the plate to the other, pieces of the plate fly off. When many cracks proceed like this simultaneously, the plate shatters. Some of the destructive energy ends up as airborne vibrations—the awful sound of breaking tableware.

Had metal been available, many of the early potters may have become smiths instead. Unlike pottery, metal is shiny, reflective, and smooth. Often cool to the touch, it doesn't break so easily. Its internal anatomy is not quite so rigid as the inside of a ceramic material. Most familiar metals dent and bend before they shatter. They are strong both in compression and in tension, which is why steel cables and beams can hold up massive structures like the Golden Gate Bridge and the Empire State Building. Heat metal and you often can beat it into sheets, foil, or, in some cases, metallic leaf far finer than tissue paper. A cubic inch of heated platinum can be drawn into a wire so fine that it would wrap around the world twice—more than 52,000 miles. Heat metals some more and they melt, which means you can pour metal into molds and cast all kinds of intricate shapes out of them.

The seeds for the Ages of Bronze and Iron, much like those for the New Stone Age, were always lying around riverbeds, in caves, and in the bush. During their thousands of wandering generations, even the most ancient of Paleolithic hominids must have happened upon eye-catching nuggets of copper, silver, and gold. Some may even have found meteorites rich with a nickel-iron alloy, the sort of metallic mix that later became the material backbone of an industrial revolution, even of superalloy jet engine parts.

Like the Venus of Vestonice, the first use of metal was likely confined to embellishment or ceremony. One of the oldest known metal objects worked by human hands is a tubular copper bead, probably part of what was at the time an exotic object—a necklace made out of metal.

Metal remained a rare sight in Neolithic times because there was not a lot of it just lying around ready to use and people of that epoch had no clue that rocks—even rough, dull ones—harbored massive supplies of shiny, lustrous metals whose properties are as different from stone as stone's are from wood and flesh.

The discovery of smelting metal from ore most likely was a fateful accident. Perhaps it occurred like this: while warming themselves by a fire on a night with strong, steady winds that could intensify the flames, some Neolithic herders may have noticed that a searing reddish liquid was bleeding out from the center of an impromptu cooking furnace that they built using some greenish malachite, a common copper-bearing rock. The combination of fiery heat and the carbon from the burning fuel was enough to liberate the metal hidden in the green stone.

Unbeknownst to these inadvertent smelters, carbon atoms were infiltrating microscopic pores and openings in the rock and absconding with oxygen atoms that used to hold the metal atoms in a vicelike oxide grip. Copper atoms plus oxygen atoms become copper oxide molecular structures. In sufficient quantity such molecular structures will scale into tiny grains that geologic processes can coalesce into a rocky hunk of ore. Oxide-based copper ores minus oxygen equals copper. Sulfide-based copper ores minus sulfur also can equal metal, but not without a rotten-egg odor. Little red rivulets of molten copper might have hardened into the contours of the ground before the herders' eyes. If the clay-to-pottery transformation was magnificent, the ore-to-metal transformation was divine. To smelt copper from stone was to make stone bleed.

So dramatic is the change from ore to metal that people around the globe attached spiritual, magical, and religious significance to it. If rock could become metal, then why could not copper or lead become gold? The alchemical quest was born. Well into the eighteenth century this magical component of metallurgy beguiled even Isaac Newton, one of the greatest scientists and most learned alchemists of all time. Newton wrote over one million words on the topic of alchemy, whose practitioners sought, among other things, to transform less magnificent materials such as lead into what many people still consider to be the most magnificent material of all, gold.[5]

The initial era of metallurgical discovery unfolded between eight thousand and four thousand years ago. Some of the earliest known metallurgists lived in the Zagros Mountains, a desolate and rocky area stretching from what is now western and central Turkey to the edge of the central desert of Iran. The land there was not so suitable for agriculture, but it was rich with metallic ores.

The use of metallic materials bloomed into diversity in much the same way that pottery had. People in various places worked with various ores and smelting techniques. The early metallurgists must have been a doughty but sickly breed. They had to endure the dark and stale air of mines; they breathed the poisonous and foul-smelling fumes of roasting ores; and they withstood searing heats from sun-bright furnaces and the back-breaking labor of turning raw metal into useful and pretty implements.

"With the advent of smelting, feverish experimentation with ores, woods, and charcoals, and with furnace designs, blowing devices, and clays, was carried out," is how one historian imagines metal technology to have swept the world.[6] Cyril Smith envisioned this fever to have played out this way: "Practically everything about metals and alloys that could have been discovered with the use of recognizable minerals and charcoal fires was discovered and put to use at least a millennium before the philosophers of classical Greece began to point the way toward an explanation of them."[7] In other words, metallurgical innovation took place well before anyone had ever thought that all materials, including metals, are made of atoms.

The copper beginnings of metallurgy did not raise the technological stakes quite enough to inaugurate an historical age the way bronze and iron metallurgy eventually did. Copper is soft and malleable, fine for ornaments and vessels. It is not so good for weapons or for making rugged tools. In a mold of fired clay or stone, molten copper shrinks as it cools, and it absorbs gas, which is why cast copper objects often have pits, bubbles, and other imperfections that weaken the product. Moreover, copper ore was not so easily found in the days before mining was common. Its rarity initially made it valuable and expensive, the stuff of the rich and famous.

Red copper combined with whitish tin, an even softer metal, yields yellowish bronze. Bronze is a metallic alloy that is unlike either of its components: harder than copper or tin, it can hold shapes better than either component, including sharp edges. Bronze alloys, which melt at temperatures between those of proper copper and pure tin, are also easier to work with than the alloys' ingredient metals. Molten bronze flows more readily than copper alone, so it casts more successfully. A material that casts well and can hold an edge is a material able to equip armies with harder, sharper, and deadlier weaponry.

The original discovery of bronze was probably yet another well-exploited accident. Early copper workers could well have hauled back to their smelting hearths copper ore that included some tin-rich ore, perhaps the brownish-blackish cassiterite.[8] Astute observation eventually could have yielded the following rough rule for smelting bronze: add one part of black ore to ten or more parts of the green ore, and then put the mixture into a draft-blown fire. The result was a new material that for the first time was strong and tough enough to replace many uses of stone, which persisted as a staple of the tool chest.

Because tin-bearing ore was even scarcer than copper ore, bronze initially was rare and therefore went into making priority implements: weapons. It was not a material for a commoner. For its pivotal role in making bronze, tin became a militarily and politically strategic material. To secure supplies for their military metallurgists, Greek and Phoenician seamen as long as four thousand years ago undertook long sea voyages to the southern coast of Britain, where the ground was rich with tin ore.[9]

Roughly at the same time, Chinese metallurgists were doing the world's most sophisticated bronze work.[10] Their bronze often contained 15 percent tin and 85 percent copper, the best proportion for casting bronze: it produces an alloy three times harder than copper. Starting with the ore itself, precision mixing of the components is nearly impossible since the metal content of even two pieces of the same ore can vary greatly. So the Chinese bronze workers must have discovered their 15 percent solution by smelting copper and tin separately and then melting the two metals together in many different ratios until they found a ratio that worked best—trial and error at its best. With their optimal bronze alloys they cast beautiful objects like ceremonial bells with ornate designs and inscriptions first imprinted as negatives into the fired clay surfaces of the casting molds. One pyrotechnical product, pottery, had become an important means of processing another—metal.

As more sources of metal were found and as mining became more widespread, bronze eventually became more commonly available, though rarely in high volumes. As a class of materials, metal remained generally exotic. Most people were surrounded by things that could be made without using fire. Wood, stone, mud, thatch, hide, wool: these were the textures that filled the landscape.

Iron was the first metal for the masses. As far back as five thousand years ago, metalworkers must have somehow gotten their furnaces especially hot on a day when they had charged the furnace with some iron-rich rock, maybe blood-red hematite. Perhaps it was a consistent blast of salty wind sweeping into their furnace from the sea that made the difference. From this they obtained a silver-gray metal that they had never seen before. It was iron. One of the oldest known iron implements is a dagger blade from 1350 B.C. It must have been a highly prized item because it was found buried in the mummy wrappings of the entombed Egyptian pharaoh Tutankhamen.[11]

Iron eventually became everyman's metal because its ore is widespread and plentiful: more soldiers could be equipped with it than with bronze, and iron tools were more readily available to farmers, potters, clothes makers, and other workers. Iron democratized metal.

The downside of iron is that wrenching it from ore requires a temperature hundreds of degrees higher than that needed to extract copper or tin. In fact, the melting temperature was too high for ancient furnaces, so early iron-workers got their metal by beating it out of a solid "bloom," a conglomeration of metal and rocky slag that forms when the ore is heated to temperatures high enough to loosen things up in the ore but not high enough to melt iron. Only through sweaty, exhausting ear-wracking work with hammers could a smith consolidate the bits of metal that would form in the bloom like raisins in a rocky

Steel

pudding. It was then the blacksmith's job to reheat and beat the consolidated iron into useful things.

The earliest iron turned out no better than bronze. Often it was worse, especially for weaponry. Compared to copper and bronze, it was a slightly crazy metal, with unpredictably varying qualities, so a smith never quite knew what to expect from a finished iron implement or weapon—often they came out too soft or too brittle.

The reason for this fickleness is the special role that carbon atoms play in the properties of iron-based metals, including steel. No one knew in the early days of iron that carbon atoms from the burning fuel in their bloomery and smithing furnaces were actually becoming part of the metal's internal anatomy and changing its properties. In the smelting of copper or tin, carbon's role is to remove oxygen without remaining in the metal. The iron that smiths wrought from the blooms and then formed into weapons and implements absorbed varying amounts of carbon. Too little carbon resulted in a consistency closer to that of pure iron, which is soft and malleable. A man wielding a pure iron sword would be no match for an opponent swinging a bronze sword. Too much carbon in the iron, however, yielded a brittle metal that could shatter like pottery.

Most often smiths ended up with iron containing somewhere around 1 to 3 percent carbon by weight. When lucky, they would get somewhere between .1 and 1 percent carbon. When that happened, the metal underwent a dramatic personality change. It became much harder and tougher, and it could hold a much sharper edge. When a smith managed to keep the carbon content within this range, he made steel.

By around 1500 B.C. the Hittites of Asia Minor and Syria had developed a sizable iron industry.[12] They were making enough to equip entire armies, not just their elite commanders, with iron weaponry of renowned superiority. Perhaps their smiths figured out how to convert iron consistently into steel.

Steel or not, their metallurgical advantage translated into military success.

Much later, by about the seventh century A.D., Samurai sword makers in Japan did master the art of steelmaking, as had their counterparts in Damascus, Syria.[13] Their success was passed down from master to apprentice in precise, secret rituals that were the precursors of the exact physical and chemical measurements and process control with which modern steel producers insure uniformity in their products.

These sporadic success stories notwithstanding, steelmaking remained an elusive and tricky art. Steel did not emerge from its status as a specialty small-volume metal until the middle of the nineteenth century, when it first became available in tons rather than in pounds.

Aside from turning clay into pottery and wrenching metal from rock, fire could also transform mineral sand into glass, the third of the ancient trinity of materials that were born of flame.

Obsidian, nature's own volcano-made glass, had long been a coveted material, probably well back into the Paleolithic age. Natural glass forms when volcanic heat melts silica-based rocks into liquid that then cools too rapidly for the atomic constituents to reorganize into the tiny crystal grains that comprise a lot of stone and rock. There are too many light-absorbing metal atoms and too many light-scattering defects in a piece of obsidian for it to be transparent. Instead it is translucent and usually of a dark-brownish or green hue. No other material could be made as sharp-edged as obsidian. But it was relatively rare, and its edges were easily chipped and broken.

The first people to emulate the volcanic invention of obsidian to make glassy materials using fire probably lived somewhere in Egypt about six thousand years ago, in the heyday of copper. Atop small pottery beads they applied a coating of colored mineral dust. The heat of the fire then melted the dust particles enough for them to fuse into an unbroken and uncommonly smooth coating—a shiny and colorful glaze.

Among the oldest entirely glass objects known are opaque glass beads, probably 4,500 years old, that were found in Iraq (then part of Mesopotamia) and elsewhere in the Near East.[14] In the days of the pharaohs, Egyptians and Mesopotamians had developed a type of glass craft in which they wound taffylike coils of hot glass around a clay core. They then chipped out the core to leave behind hollow and beautiful glass vessels.

The exact circumstances of the original discovery of glassmaking recedes into the same historical mist that shrouds the discovery of pottery and metal extraction. There are only stories like this one from Pliny, a first-century Roman historian. He credits Phoenician sailors with the discovery. As he had heard it, the sailors landed on a sandy beach, where they built a cooking fire. To heat a cooking pot, they propped it up on some blocks of natron, an alkaline material derived from the ash of plants (and from which the chemical symbol for sodium is derived). The sailors found that the combination of heat, sand, and natron produced a hot liquid stream of "an unknown translucent liquid" that cooled and solidified into glass.[15] As in glassmaking today, an ingredient akin to natron serves as a flux that sops up impurities and lowers the melting temperature of the mix.

Whatever its origin, glass was not a substance for the masses. Like gold, the sky-blue lapis lazuli, and translucent-white alabaster, glass was the stuff of Pharaohs, high priests, and nobles. What helped bring glass into the realm of common materials was the invention of glassblowing by craftsman in Syria and Palestine about two thousand years ago. By blowing air into a dollop of molten glass gathered at the end of an iron blow pipe, the workers could make beautiful, hollow glass vessels. Within a hundred years Roman glassmaking was big-time commerce, even selling glass "gladiator-beakers" for patrons of the games. These were ancient souvenir cups with imprinted scenes of gladiatorial battles transferred from the ceramic molds into which the Romans blew the molten glass.[16]

The architectural potential of glass's translucence was not realized until about the twelfth century with the construction of Gothic cathedrals. Intensely colored, backlit portrayals of biblical stories bathed praying congregations, enhancing the sacral magnificence of the massive stone cathedrals. In the following centuries transparent glass made from purer silica-based ingredients began replacing the dark wooden shutters and translucent oiled paper or muslin that kept the climatic elements out of the typical home. By the mid-fifteenth century, half of the houses in Venice, Europe's most influential glassmaking center, had glass windows. Two and a half centuries later, the transition to glass in Europe at last was nearly complete.[17]

Some historians credit the added illumination made possible by glass with heightening interest in cleanliness and hygiene. Windows made dirt more visible. And thanks to superior mirrors— made with transparent glass that reflected properly from the thin metal foil on one side—people could see and understand themselves and their conditions more accurately than ever before; glass, a miraculous substance that is at once as solid as a rock and as invisible as air, shed as much light on people's minds as on their surroundings.[18] Moreover, the magnifying powers of glass eventually enlightened scientists as well, enabling them to understand what it is inside of materials that makes the stuff of the world the way it is.

By sheer volume it was another less lofty set of pyrotechnical materials—cement, concrete, plaster, brick, and mortar—that changed the look of the world for ordinary people far more than metal, pottery, and glass. Like all materials used by people, the arrival of each of these had its own discoverers or inventors.

Plaster, for example, goes back at least eight thousand years. In what was Anau, Iran, a plaster based on gypsum (a crumbly mineral that when pulverized and mixed with water can first be spread before it hardens) was becoming, as one historian has put it, "a trademark of the Iranian plateau that has lasted right into the twentieth century."[19] Likewise lime, or calcium oxide, a powder made by burning limestone and that would much later become a primary ingredient in steelmaking, was formed into an early plaster that ended up on floors in ancient Turkey. In the sixth century B.C. bricks, thousands upon thousands of them, were becoming the architectural trademark of Babylon, one of the great cities of antiquity. The builders of Babylon, and of the rest of ancient Mesopotamia, formed wet clay into bricks and then fired them to hardness in kilns. Rome, too, became a city of brick after much of the city burned to the ground in A.D. 14. After that, brick became a universal building material for urban residential structures as well as for imperial buildings.

The Romans pioneered the use of hydraulic, or water-cured, cement. Its unique chemical and physical properties produced a material so lasting that it stands today in magnificent structures like the Pantheon. Yet the formula was forgotten in the first few centuries after the fall of the Roman Empire and wasn't rediscovered until 1824 as Portland cement. One Roman version was based on a burned mixture of two major components: volcanic ash—called *pozzolana*—from Mount Vesuvius, which destroyed Pompeii and nearby towns in A.D. 79; and calcium carbonate, the stuff of seashells, chalk, and limestone.[20] Adding water to these sets off a complex set of chemical reactions that convert the gritty pasty stuff into what is essentially artificial stone. The nineteenth-century rediscovery of Roman cement, the aforementioned Portland cement, is made from a combination of burned limestone, clay, and water. It is the single most heavily used human-made material on earth. It is visible everywhere as the flesh of skyscrapers, dams, bridges, roads, patios, and thousands of other structures.

In addition to the hard materials made from inorganic or

mineral sources—pottery, metal, glass, plaster, bricks, cement—new soft materials were also finding their way into the materials pantry. This history of paper started unfolding in about 3000 B.C. Ancient Egyptians found that they could pound thin bark peeled from river reeds—papyrus—into sheets. It was easier to write on the papyrus sheets than on stone or baked clay tablets, and the result was easier to carry around.

Besides papyrus, ancient material tinkerers found ways of mashing, pressing, heating, filtering, and distilling vegetable and animal tissues to get substances that were tarry, waxy, sticky, and gluey. There are plenty of biblical references to materials that served as perfumes and ointments. Other historical evidence from ancient shipwrecks and archaeological sites has revealed the remains of salves, dyes, glues, waterproofing agents, possibly even antiseptic chewing gum.

By Christ's time the diversity of materials made and used by the world's modest population would have astounded George, the Paleolithic inventor of the first material technology. The transformative power of fire was raging throughout the globe as people mixed it with earth, air, and water to create the many substances of their constructed landscapes.

TWO

NEW *Stuff* IN THE OLD WORLD

In the nineteenth century almost all of the materials people were making and using had been made and used for several thousand years. But between ancient times and the 1800s, the scale of production had been rising. Animal-, wind-, water-, steam-, and finally electric-powered machinery had been put to ever increasing use to keep up with the accelerating demand for materials from growing populations. More and more mines opened. Smelters, potters, glassmakers, metallurgists, and other materials workers were fruitful and multiplied. Trade in materials and goods brought the world's people into commercial bonds. Production of materials was so intense at times that laws were intermittently passed to prevent entire forests from disappearing in furnaces and kilns.

Craftsmanship with materials had long become a matter of national pride. So coveted were the secrets of the Venetian glass houses, which were sequestered in 1291 on the island of Murano within sight of Venice, that workers were forbidden to leave the island. Attempted escape was punishable by execution.[1]

For some the passion for certain materials was strong enough to inspire them to try and find out what makes materials tick, to look inside of inanimate matter in the hope of finding a key to their creation. Augustus the Strong, King of Poland and monarch of Saxony (located in what is now southeast Germany) from 1694 to 1735, expressed one of the more illustrious examples of this curiosity. And it presaged the kind of business thinking that would become common in the twentieth century at places like DuPont and General Electric. As the materials scientist and historian David Kingery of the University of Arizona has put it, "He and his court supported one of the first truly modern research efforts: it was aimed at porcelain manufacture."[2]

The monarch had a passion for the beautiful Chinese and Japanese porcelains that were flooding European markets in his day. Ownership of these ceramic wares was a mark of social stature. No one in Europe had a more extensive collection than Augustus, but his interest was not merely aesthetic: he wanted to create a domestic source of comparable quality.

To this end he commissioned Count Ehrenfried Walther von Tschirnhaus, an aristocratic Bohemian trained in mathematics and physics. He was known for mineral-melting experiments in which he had used not fuel-fed furnaces but rather "burning mirrors" made of high-quality glass lenses to generate the high temperatures required for the process.

Tschirnhaus first traveled throughout Europe to see what others had learned. He then set up a secret laboratory in Meissen, where a hired staff conducted experiments on many combinations and preparations of different clays and clay mixtures. Eventually, Tschirnhaus's chief researcher, the alchemist Johann Friedrich Böttger, pursued a promising lead. He found that when combined, silica (the stuff of sand, flint, and certain stones, including quartz) and lime (calcium oxide made by burning limestone) melted at a lower temperature than either ingredient alone. The cooled and dried product appeared whiter and finer than previous combinations.

Material science has since explained the reason for this result: the lime serves as a sponge, or a flux, for collecting contaminants in the silica that otherwise degraded the whiteness, translucency, and texture of fired wares. The melting temperature of the mixture was still high enough (more than 1,350°C) that the Saxon researchers needed hotter burning furnaces than current furnace designs could muster. So they devised them.

The effort paid off. Tschirnhaus and Böttger developed a new European porcelain with the "white body, translucency, and sonorous ring of the Chinese wares,"[3] as Kingery, whose students repeated some of the Meissen experiments, describes them. The ceramic wares became famous, coveted and valuable throughout Europe. "The composition and microstructure are similar to some of the spark plug porcelains developed during the early years of high-compression automobile and aircraft engines," Kingery notes.

Another pioneer of the early eighteenth century, Josiah Wedgewood, whose name is still synonymous with world-class ceramic ware, followed the Saxony example of systematic process development. One of his innovations, developed in the early 1720s, comprised a series of ceramic cones made of standardized clay mixtures. By measuring the shrinkage of these cones as the furnace temperature increased, he had devised a crudely calibrated but effective pyrometer for measuring the high temperatures inside a furnace.[4] The measurements, in turn, helped him attain an enviable level of quality control. Before that, temperature control in a furnace had been entirely a matter of professional judgment, typically involving the withdrawal and examination of samples from the furnaces during firings. It was like baking a cake in an oven with no temperature dial.

Neither Böttger nor Wedgewood knew exactly how, but their careful process control was working because it was coercing the inner anatomy of their products into favorable material structures. Ultimately what makes steel steel and porcelain

porcelain is the way their constituent atoms and molecules interact with one another and aggregate into ever larger structures until they form something you can hold in your hand and use. It is like the progression from atoms to molecules to organelles to cells to tissues to organs to entire animals. This multilevel, hierarchical anatomy ultimately determines the properties of every person and every material. Getting control over this hierarchical structure always has been the key to making consistently good materials.

In the 1720s a French contemporary of Wedgewood's, the chemist René Antoine Ferchault de Réaumur, reasoned his way to perhaps the most prescient foreshadowing of this modern picture of material structure. He conjectured that steel was made of an arrangement of tiny grains that in turn had a substructure made of parts that he called "molecules," though his concept does not correspond to the modern definition of a molecule as a specific arrangement of atoms. His "molecules" had their own substructure, some kind of less-than-crystalline jumble of spheres. On the finest scale he imagined a periodic, or crystalline, stacking of even finer spheres. When modern materials scientists see the diagram of nested structures that Réaumur prepared in the 1720s to illustrate his ideas, they say, "Not bad!" (See Figure 2.)

Still, even for later generations of scientists in the nineteenth century who knew that the properties of materials and the way they performed depended upon their inner structures, there wasn't much of a foundation for probing very far into material anatomy. After all, the very existence of atoms, much less how they might assemble to become this or that material, was a hotly debated topic until the very end of the century. Moreover, there were no laboratory instruments for peering deep enough into the anatomy of materials to know with any precision what the atoms and molecules actually were doing or how they were arranged. These limitations, however, did not hold up the beginnings of a science of materials.

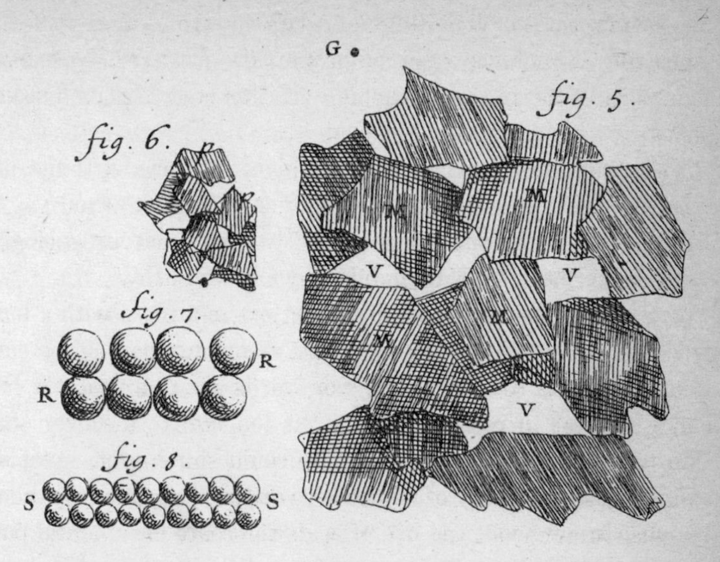

FIGURE 2

Réaumur's Prescient Picture of Stuff. René Antoine Ferchault de Réaumur, a French chemist of the eighteenth century, envisioned the internal structure of metals as a series of nested architectural levels. This 1722 illustration of Réaumur's view of the internal structure of steel is remarkably like the generic portrayal of materials by modern materials scientists. The dot marked *G* represents a grain of steel as seen by the eye. Below is the grain's structure as seen under an optical microscope. The structure includes smaller subgrains (*M*) and spaces. Below this level, Réaumur surmised that each subgrain was itself composed of yet smaller subgrains (*P*) and voids with the series, perhaps continuing for several additional levels.

THE ORIGINAL IMAGE BY RÉAUMUR IS REPRODUCED IN CYRIL SMITH'S *HISTORY OF METALLOGRAPHY* AND IS REPRINTED HERE WITH PERMISSION FROM THE MIT PRESS

The challenge was daunting. Even knowing about atoms would barely be a beginning. Trying to understand materials by way of atoms and molecules would be like trying to understand how a society works by looking only at its members individually. Understanding societies requires insight into the dynamics of social interaction. It is the same for materials. Atoms and molecules invariably are a material's fundamental ingredients, but they are part of chemical and physical features and interactions over a range of architectural levels that really constitute a material's set of properties. Any real science of materials would have to grapple with this complexity on scales that range from atoms to bridges.

A founding father of modern materials science came not from the ranks of nineteenth-century chemists with their vague notions of atoms and molecules; he was an amateur geologist working in Sheffield, England: Henry Clifton Sorby.[5]

Sorby came from a family of cutlers in a town with a long tradition in metallurgy. Metals were as much a part of his context as wood is to a carpenter's or words are to a writer's. His early interest in rocks and minerals led him to discover that thin flat sections, through which he could shine light, revealed enough gross anatomy of minerals to distinguish amongst them. He also knew about the use of acids to create ornamental patterns on metal items such as razors or to prepare metal plates used for illustrations in catalogs. In the heat of the summer of 1863, Sorby combined these two seemingly disparate lines—slicing stone and etching with acid—into a technique by which he was able to show the microstructure of iron and steel as it had never been seen before.

The process became known as metallography because it was like writing on metal. It starts with a cut and polished surface of a metal specimen, followed by a gentle etching step that uses dilute nitric acid. Under a microscope Sorby could see the result as a puzzlelike montage with regions of different geometries and appearances that he was able to relate to the process by which the metal was made, to the material's starting ingredients, and to the materials' strength, fracture resistance, and other engineering properties. Though Sorby's techniques didn't go much beyond his own home at first, by the late nineteenth century metallography had became a standard method for studying metals and for controlling their quality. It still is.

The technique of metallography provided the kind of evidence needed for understanding how to produce iron or steel of consistently high quality. Armed with that kind of knowledge, materials research stood on the brink of the modern era; rather than shooting in the dark, it could now turn the light on first.

Sorby's work represented the beginnings of a genuine science of materials in which a material's composition, internal structure, and method of preparation are systematically related to the material's properties and technological performance. Even today, these interrelationships comprise the framework of modern materials researchers.[6] But Sorby's metallography was only a key to a well-secured mansion. To get beyond the foyer would take many more keys.

The incentives for looking more deeply into the anatomy of materials were intensifying in Sorby's day. Iron and steel already had become materials of strategic military, economic, technological, and social importance. Iron and steel were the lifeblood of the Industrial Revolution, which began in the eighteenth century. Building more and better metal was the key to better railroads, stronger armaments, and more reliable steam engines to power the heavy machinery in factories and mills.

The production of raw iron (pig iron) in England, the world's biggest producer throughout much of the eighteenth and nineteenth centuries, mushroomed between 1740 and 1850 from under 20,000 tons to about 2.5 million tons.[7] Abraham Darby, the third generation of a famous family of ironmasters in Shropshire, England, gets credit for the first big iron structure ever—the aptly named Iron Bridge. It opened in 1779 over the River Severn in Shropshire. Steel, even though its basic composition is almost all iron with a small dose of carbon, remained expensive to make, even after 1850, when English smelters were producing about 60,000 tons per year in small batches that a single man could lift. Even when pooled, that's enough steel for at most a handful of modern skyscrapers.

In their new technological roles, iron and steel had to perform under conditions more stressful than metals had ever faced before. Never did materials have to withstand such high heats and pressures for so long as when they became the stuff of steam engines. Never did materials have to endure the clacketa-clacketa, metal-against-metal stresses that go with trains on rails. Never before did materials have to be as predictable, stable, and standardized as they did in mass manufacturing using interchangeable parts. When metal in the form of rails, wheels, pistons, boilers, and gun parts failed, there were fatal explosions and derailments. Death on the rails made newspaper headlines far more frequently in the nineteenth century and early twentieth century than it does today.

The physical limitations of cast-iron cannons in the Crimean War (1854–1856)—in which England, France, Turkey, and Sardinia defeated Russia for the domination of southeastern Europe—were a turning point in the history of steel. Those limitations caught the attention of Henry Bessemer, an Englishman who already had enjoyed considerable commercial success for improving the centuries-old pencil by replacing its "lead" with a preparation containing powdered graphite, a carbon mineral whose internal structure is characterized by loosely bound molecular sheets that can slide off easily to mark paper. But high-volume steel production was about to become his more enduring claim to fame.

Bessemer had developed a new kind of projectile that could travel farther than cannon balls because it developed a spin as it left the cannon barrel. But the new projectile, which he patented in 1854, also generated more stress on the cannon's iron. Napoleon III, the French Emperor, was interested and bankrolled Bessemer to work on the projectiles. In December of 1854, Bessemer successfully demonstrated that an elongated projectile could be made to spin when shot out of a smooth-bore gun, but the French commandant who witnessed the trial questioned whether the cast-iron guns of the day could withstand the added stress that the new projectiles imposed. In his autobiography Bessemer wrote, with a characteristic lack of humility, that "this simple observation was the spark which kindled one of the greatest industrial revolutions that the present century has to record."

The standard method of steel production at the time was hopelessly small in scale and variable in result. A charge of wrought iron—itself the product of time-consuming labor—was sealed in a clay crucible with some charcoal and then heated for up to several days. If everything turned out right, the amount of carbon needed to make a steel alloy would wend its way into the metal to produce about 50 pounds of steel over the course of ten days. This laborious method precluded large-scale steel production for war materiel like cannon and ordnance or for big structures like bridges and buildings.

Bessemer's breakthrough in 1855 was to drop the standard approach, which involved making steel by heating carbon-free wrought iron with carbon-bearing fuel. He tried to turn the process around by starting with carbon-rich pig iron and using oxygen in an air blast to get the excess carbon out.

"Pig" iron, the raw product of iron smelting, got its name because molten iron often drained from smelting furnaces down central gullies and then right and left into rectangular molds where the molten metal cooled and solidified. The arrangement looked like a bunch of suckling pigs. Along with a carbon content of about 4 percent by weight, pig iron also contained impurities that varied with the type and quality of ore. To turn pig iron into steel, about three out of every four carbon atoms somehow would have to be removed.

That sounds like a hard task for solid pig iron. In the molten state, however, oxygen passing through would readily grab onto carbon atoms. Moreover, the oxygen would bind at least some of the impurities present in the melt into oxides that would form a rocky slag. Slag is less dense than molten metal, so it floats to the top, where it is readily skimmed off. What's left in the melting crucible, called a converter, is purer metal.

The combination of oxygen and carbon during the blast produced a carbon monoxide gas that shot up and out of the molten metal, spewing white-hot flame and sparks like an artificial volcano. As expected, the oxygen in the air blast did form oxides with the silicon, manganese, and some of the other impurities that he wanted to remove. The process also liberated an unexpected amount of heat that displaced the need of fuel for keeping the metal in a molten state.

There was far more to the spectacle than sparks and flames. The combined advantages of reduced impurities, lower carbon content, and the heat generation with no-cost air (as opposed to costly fuel) led to more than just better iron. As the carbon left the pig iron, the pig iron became more and more like steel. If the air blast lasted about twenty minutes, the amount of carbon left in the iron could produce a "mild" steel that would be as workable as a carbon-free wrought iron yet much stronger and tougher. A cannon made out of good steel will not shatter as easily as an iron cannon.

After months developing machinery that would enable workers to operate the blowing engines that created the air blasts for the converters, Bessemer publicly introduced his iron-to-steel conversion process on August 12, 1856. The paper he read to the British Association for the Advancement of Science that day was published in its entirety two days later in the *London Times*.

That the process at that point was far from perfect became apparent to the many ironmasters who licensed Bessemer's patent. The ores of the Swedish pig iron that Bessemer used for his earlier oxygen-blast experiments were essentially devoid of phosphorus. But the domestic supply of pig iron had plenty of phosphorus, and all it yielded at the end of a Bessemer conversion was an inferior batch of metal that would crack during attempts to work it into implements. The phosphorus would always end up between the metal's individual grains, weakening the bonds holding the metal together. It looked as though Bessemer's fifteen minutes of fame was up.

Bessemer eventually decided to make his process work by

relying on imported Swedish iron. It was a pair of cousins, Sydney Gilchrist Thomas and Percy Carlyle Gilchrist, who in 1875 found the way to disarm the phosphorus conundrum by adding limestone both to the firebricks of the converter and to the iron charge inside the converter. The phosphorus reacted with the calcium oxide of the burned limestone to form a slag that could be skimmed from the metal. That opened the Bessemer process to much more of the world's iron ore.

Another problem with the original Bessemer process derived from the oxygen. There was enough of it in each blow not only to take out the excess carbon in the pig iron but also to blow holes in the subsequent steel. This led to a flawed product known as "burnt iron." In this case Robert F. Mushet (who later would develop a tungsten-manganese steel alloy whose especial hardness extended the lives of tools made of it by five or six times) came to the rescue. He discovered that adding an alloy of manganese and iron known as spiegel to the charge would keep the oxygen under control.

With the subsequent patchwork by Thomas, Gilchrist, Mushet, and others, the Bessemer process succeeded in turning steel into a high-volume commodity. In twenty minutes a large Bessemer converter could transform 30 tons—60,000 pounds—of pig iron into steel. Global steel production throttled upward at an astounding rate: from about 500,000 tons in 1870 to nearly 28 million tons by the turn of the century,[8] by which time other metal entrepreneurs had developed additional methods for producing large amounts of steel.

The giant leap in iron and steel production delivered huge supplies of materials with an unprecedented combination of strength, resilience, hardness, and formability. There was a proliferation of steel alloys with various properties that depended on the different proportions of their constituent elements. The special toughness of nickel steels fueled a naval arms race among European powers who were using it as armor plating for their warships. More tinkering brought forth stainless steel, made with a good portion of chromium, some nickel as well as other minor constituents. An alloy like stainless steel, which was resistant to rusting and corrosion, was dramatic evidence of metallurgy's ability to bring forth spectacular new properties in basic materials. No one understood how the alloying elements went about changing the metal's properties, but empiricism has always been a fruitful, if not always efficient, means of discovery.

Iron and steel became the stuff of big and visible things—ships, bridges, buildings, even the Eiffel Tower. These metals, particularly steel, were the harbingers of skyscrapers and audacious urban skylines. In 1874 James Eads used steel to build a giant arched bridge across the Mississippi River at St. Louis.[9] Some 50,000 tons of steel went into the building of the Forth Bridge near Edinburgh, Scotland.[10] For the enormous suspension cables that would hold up his Brooklyn Bridge, John Roebling in 1883 could choose steel rather than the iron that Abraham Darby was confined to a century earlier. In 1890 the second Rand-McNally Building in Chicago was the first building to be framed entirely in steel.[11] Hundreds of ever higher buildings and skyscrapers followed. Steel also was showing up in hundreds of smaller everyday items like food cans, home appliances, automobiles, and road signs.

For all of steel's social impact, the breakthroughs that allowed mass production of it owed little to academic science. Sorby's metallography, for example, did not become part of standard practice until late in the nineteenth century—well after high-volume steel production had become commonplace.

Some rough guidance actually had been available much earlier. Even by the time Antoine Lavoisier was developing chemistry into a genuine science, chemists had learned that the difference between iron and steel pivoted on a small difference in the amount of carbon present in the metals.[12] Pig iron contains about 4 percent carbon by weight. Wrought iron contains almost none. Steel contains no more than about 2 percent, usually less. Practical steelmakers like Bessemer learned to hit the right compositions and processes by trial and error, not through chemical analysis.

That began to change in the later nineteenth century. In 1870, for example, The Iron and Steel Institute was founded to promote a more systematic and scientific approach to making and developing steel. A decade later Sir William Siemens developed a rival steelmaking process that was slower than Bessemer's but afforded more control over the steel's composition; he remarked that "a taste for science had been awakened among employers, and men who formerly ridiculed the idea of chemical analysis now speak of fractional percentages of phosphorus with great respect."[13]

Nineteenth century steelmakers were not the only ones discovering the practical power of chemistry. When Sorby was inventing metallography in the 1860s, the chemical industry was in the throes of a massive transformation from its alchemical and protochemical roots. It was evolving from a collection of mostly unrelated, small-scale producers of specific substances for tanners, glassmakers, dyers, and other craftspeople to a nexus of interdependent producers. One person's trash became another's feedstock.

As steel production rose, chemistry began to emerge from academic obscurity to become an industrial and economic force. New chemical dyes for the textile industry were the first big products, some of them quickly overtaking the traditional dye industries such as the indigo growers of India. The synthetic dyes were loud signals of chemistry's miraculous ability to transform stuff. Many of the new dyes were made of chemicals like aniline that were obtained by distilling coal tar. Coal tar was a thick, goopy, unpleasant product of the gas lighting industry that would have been a total waste product had it not found a use in preserving millions of wooden ties during the railroad-building frenzy of the 1830s and 1840s.

chemical dyes

Beautiful, colorful, big-business dyes from black, goopy waste were just one tip of one of chemistry's icebergs. With increasing frequency chemists were happening onto new chemicals. These, in turn, provided new chemicals for reacting with already known ones. This gathering combinatorial momentum spurred a dramatic growth in the catalog of chemicals and materials. The diversity of transformations emanating from chemistry labs rivaled the fantasies of any medieval alchemist. Solids became liquids, which became gases, which became solids again. Gunk became crystal pure, and crystalline purity became gunk. Over and over again in laboratory glassware, colors, textures, and forms changed with hallucinatory strangeness, degree, and surprise.

As it always had been, the chemists' raw materials were of the earth: ores, minerals, and clays, pulverized into powders; coal and wood, which, when heated without burning to the point of decomposition, yielded smoky "spirits" that would then cool and condense into tars and liquids such as turpentine; kelp, cotton, and other vegetable matter burned to yield alkaline ashes, already pivotal for millennia in metallurgy and glassmaking; and petroleum (known many centuries earlier by the Islamic alchemists in the Middle East, where it would seep up to the surface beckoning for attention), which began to serve up a chemical cornucopia of fuels, sealants, paving and construction materials, and feedstocks for polymers.

These materials went into furnaces and electrolytic baths, where heat or electricity would wield their magic or violence. In kettles and flasks, hot liquids saturated with pulverized and dissolved minerals or extracts of the vegetable world would, with sufficient cooling or evaporation, leave crystalline salts, sugars, fats, oils, and other purified substances. Complex liquid brews could be separated into purified components by distillations in which various liquid components would boil off at various temperatures and so could be collected apart from the others as they separately fled through the top of the container to a collection vessel. Kerosene, naphtha, benzene, turpentine, ether, and many other liquids were obtained from coal and petroleum this way.

From all this tinkering entirely new categories of materials were bound to emerge, among them a tacky, bouncy, compliant material called rubber. For centuries South Americans had air-cured the milky latex (sap) of a tree known as *Hevea brasiliensis* into a curiously elastic, easily shaped, and waterproof material that yielded bowls, shoe soles, and bouncy balls for games. Its South American users called it *caoutchouc,* which means "weeping wood." In 1770 Joseph Priestly, codiscoverer of oxygen, got a small sample of South American rubber from a friend. Priestly dubbed the material "rubber" when he found that it could erase pencil marks with a simple rubbing action. For the next seventy years people experimented with latex rubber for medical uses such as making bandages, stomach pumps, and ice bags and for waterproofing fabrics and leak-prone leather water hoses.

Natural rubber had major shortcomings, however. When cold, it would become brittle and crack. When hot, it grew sweaty, flimsy, and gross. In the mid-1830s rubber got an especially bad reputation in the United States, where several firms hastily hyped and marketed rubberized apparel as a new, convenient solution to rainy days—convenient, that is, in moderate climates; the hot summers and cold winters of the 1830s revealed the limitations of untreated rubber to dismayed consumers.

Charles Goodyear, then a hardware merchant in Philadelphia wholly ignorant of the chemical arts, nonetheless was convinced there had to be some way of overcoming rubber's sensitivity to the weather. He spent years mixing rubber with various compounds and solvents while subjecting the results to various degrees of heat. Everything was worth trying—vegetable oils, ink, even chicken fat.

In an apparent breakthrough in 1838, he found that bathing natural rubber in the vapors from nitric and sulphuric acids seemed to achieve the desired resistance to disintegration in the heat. The cash-strapped Goodyear even convinced the post office to offer a contract for mail bags made of the acid-cured rubber. But the first batch of mail bags revealed the Achilles heel of the process: after spending several days in a hot storage room, the rubber crumbled into pellets, and the post office withdrew its contract.[14]

No little setback like that was going to thwart the obsessed Goodyear. And in 1839 he stumbled onto a process at a hot stove in Woburn, Massachusetts, where a small rubber company had given him space to tinker. When he accidentally spilled raw latex rubber mixed with sulfur onto the stove, it was as though he were a protohuman hominid discovering the secret of chipping stone. Rubber mixed with sulfur and then heated transformed into a stable elastic material unaffected by the normal range of seasonal temperatures. He later called the process vulcanization. Vulcan, the god of volcanoes, was an appropriate name for the process, he thought, because its essential ingredients were sulfur and heat.

Goodyear didn't know it, but the sulfur bonded into chemical bridges that linked the individual rubber molecules into a huge elastic network, like a three-dimensional spiderweb, thus preventing the individual polymer molecules from sliding by one another in hot weather or from cracking apart in cold weather. By the end of 1841, Goodyear had developed a process suitable for industrial-scale production of vulcanized rubber.

England took the early lead in rubber production, just as it had in iron and steel; the output of rubber there increased 6,000 percent between 1840 and the end of the century. While most of it went into waterproofing items like clothes, shoes, roofs, and leather fire hoses, it also found major applications in shock absorption in vehicles and heavy machinery and electrical insulation in cables.

Eventually, the demand for tires elevated rubber into one of society's material mainstays. The first rubber tires were made in the 1840s. They were solid and made for bicycles. Toward the end of the century, men with last names still familiar today—John Boyd Dunlop, André and Deoudard Michelin, for example—developed pneumatic tires that automotive engineers were introducing roughly at the same time.

As the sight of rubber became more common in the later nineteenth century, chemists were learning more about its intriguing molecular makeup. It was, in fact, a polymer, a material whose constituent molecules are made of simple chemical building blocks linked into more or less complicated chains and networks. Since Paleolithic times, hominids and their descendants used naturally occurring polymers made of biochemical building blocks like amino acids and sugar molecules, which assembled into such useful materials as hair, fur, hide, horn, and sinew.

Raw rubber is also a natural polymer that comes from trees, not laboratories. It is composed of a small hydrocarbon molecule called isoprene manufactured in large quantities by the enormous collective industry of latex-producing cells in rubber trees. And like a piece of raw flint, natural rubber requires some manipulation—Goodyear's vulcanization, for example—to make it more useful.

Rubber was just a harbinger of a whole new world of polymers. A subsequent polymer, which required more extensive chemical manipulation, made a quiet and troubled commercial debut in 1869, aiming not at the needs of a vast military or industrial empire but rather at a small coterie of sportsmen who were indisposed by an apparent shortage of ivory for billiard balls. The incentive reportedly came from Pheland & Collender, a New York City firm that in 1863 had posted a $10,000 offer for patent rights to anyone who could come up with a workable substitute for ivory.[15]

John Wesley Hyatt, a printer and mechanic in Albany,

New York, rose to the occasion, taking on the challenge with a quixotic naïveté that clearly predates our current era of hyper-specialized team research. He began by fiddling with known plasticlike materials such as pressed wood pulp and gum shellac, which were then commonly combined to make dominoes, checkers, and other items that didn't have to take a beating—definitely not the right stuff for making billiard balls. A journeyman printer, Hyatt then turned to a seat-of-the-pants brand of chemistry. He mixed and matched binding substances like glues, resins, shellacs, and gums with fillers like paper, rags, linen, wood, and ivory dust. But none of these led to a new material fit for billiard balls.

He was set on the right track by (what else?) an accidental finding—a small, thumbnail-size piece of a sturdy, transparent film he happened upon in a cupboard of the print shop he was working in. The film had formed because of a spill from a bottle of collodion, a viscous, plant-derived substance that was used mostly as an early medium for holding photographic images and as a surgical dressing. At the time, printers often would coat their fingers with the stuff for protection against the shop's many hot, metallic surfaces.

The collodion that Hyatt and competitors decided to work with came from cellulose in cotton after it was first treated with strong nitric acid to produce a nitrated cellulose whose explosive qualities earned it the name guncotton. Hyatt found that he could dissolve the nitrated cellulose in organic solvents like ether (an early and dangerous general anesthetic) and ethyl alcohol (the kind of alcohol in liquor), themselves products of the ever more diversified chemical industry. The initial result was a clear, syrupy liquid that dried as a thin, clear film.

The material's main use throughout the 1850s and 1860s was as a photographic emulsion. When dissolved and mixed with silver chloride crystals, it could be spread into a thin film onto a glass plate that then would have the curious ability to capture a photographic image. Light hitting the crystals would cause the silver atoms to clump into dark spots—the black of black-and-white photography. But could collodion be used for other things?

By trial and error, Hyatt hit upon the formula and process for "celluloid," the first commercially successful plastic. Isaiah, his brother and business partner, named it after the cellulose from which it originally was derived. The cellulose, first converted into a nitrated form by soaking it in nitric acid, was ground up and dissolved in water to form a fine pulp. At this stage Hyatt found that he could add pigments and dyes (also products of the infant chemical industry). Next he added ground gum-camphor to the mix in a ratio of one to two, and then pressed and strained all of the remaining water out. This preparation went into a mold and then into an oven heated to between 150–300°F; it was then subjected to high pressure. In his 1869 patent Hyatt wrote that the result is a solid material "about the consistency of sole-leather, but which subsequently becomes as hard as horn or bone by the evaporation of camphor."[16] It had the formability of wet clay yet was tougher and more resistant to breakage.

A year after the patent, John and Isaiah started the Albany Dental Plate Company to manufacture and sell the first celluloid products—substitutes for the hard rubber dental plates that were most common at the time. The celluloid could be color-matched more easily than the rubber, but it, too, had drawbacks: it warped easily and had a camphorous flavor, so it never seriously threatened the rubber standard.

In 1871 the brothers established the Celluloid Manufacturing Company, a name designed to reflect their ambitions beyond dental plates. The company supplied the material in more generic forms—rods, sheets, and tubes—with the idea that other manufacturers would buy these and fashion them into a huge variety of finished products. On short order the

Hyatt brothers had convinced financial backers that celluloid was a good risk, and with money from financiers, the company relocated to downtown Newark, New Jersey, in 1872.

Despite celluloid's tendency to ignite (the Newark factory burned to the ground on September 8, 1875), celluloid infiltrated many markets over the ensuing two decades. With tortoise-shell and mother-of-pearl coloring, it was a cost-effective substitute for the more expensive materials in items in the personal hygiene market such as combs, mirrors, and brushes. With coloring and texturing it served as a substitute for ivory and even for fine linen in the form of celluloid cuffs and collars.

Celluloid even ended up circling back toward its collodion roots when George Eastman and a chemist colleague, Henry Reichenback, modified Hyatt's celluloid into a new photographic film. Unlike collodion, which had to be painted onto glass plates to work as a photographic emulsion, this product could be successfully hardened and then rolled without breaking and flaking. In the summer of 1889, the first plastic film went on sale, launching the era of Eastman Kodak's load-and-shoot camera.

Besides Hyatt, many others jumped into the emerging fray of new plastic materials. One of Hyatt's competitors, Alexander Parkes, an English metallurgist and chemist, had long felt that industrialists could benefit from a widely available material that, in his words, "combined the properties of ivory, tortoise shell, [and] horn" but that could be worked into many forms, as wood and metal could. Patents were issued for various plastic materials for uses ranging from waterproofing to varnishes and lacquers to imitation bearskins to dental plates.

Ironically, celluloid never did serve its original purpose: substituting for ivory billiard balls. Its elasticity differed too drastically from that of ivory. A game with celluloid balls just felt too different. It was a sobering lesson in the intricacies of trying to substitute new materials for traditional ones. One plastic did, in fact, become the first acceptable replacement for ivory in billiard balls. Known as Bakelite, it was not so humbly named after Leo Henrik Baekeland, a flamboyant, Belgium-born chemist who invented it by reacting two simple compounds derived from the chemical gemische of coal tar near the turn of the century and introduced it commercially in 1909.

Notwithstanding the success of Bakelite, people came to see celluloid as a cheap substitute for the real McCoy, according to the historian Robert Friedel. Although it did prove itself to be a superior (though uncomfortably inflammable) material for photography, celluloid's original stigma as a cheap impostor still clings to the lab-made polymers that have followed it in the still-unfolding Age of Plastics. Yet, celluloid was the seed of an enormous class of materials—artificial polymers—that has led the way, for better or for worse, in altering the texture of the human environment in the twentieth century.

One of the early uses for polymers was as insulating material for the nascent electrical industry of the 1870s and 1880s. Rubber and other tree-made materials such as gutta-percha were among the first polymers used in this way. Bakelite found ready use in that application as well. Besides powering the growing population of electrical lights and machines, the availability of cheap electricity helped to spawn large quantities of a new metal: aluminum.

Aluminum, first isolated from its ore in tiny amounts in 1825, is lightweight, akin to silver in color, yet resistant to tarnishing and corrosion. Early on the press heralded the new metal as holding the potential to transform industry and daily life, if only a cheap way to produce it could be found.[17]

Until the 1880s aluminum was so rare and precious that it was used chiefly for high-end items obtainable only by the affluent: opera glasses and jewelry, for example, were often made with the new material, and Napoleon III of France, an early patron of Bessemer, had a set of tableware made out of it. While mills were springing up around the world to produce steel in tonnage quantities, lone technicians in laboratories were coaxing aluminum from its ore ounce by precious ounce. When a 100-ounce apex of aluminum was placed atop the Washington monument with great ceremony in 1884, it sold for about one dollar per ounce (sixteen dollars per pound). That was about the same as silver, which then was thought of as a precious metal. Besides having a fitting status for the national monument, the metal's conductivity made it suitable for the lightning-rod assembly that it was part of.

Aluminum, the third most common element in the earth's crust, was first isolated as the great proliferation of machines and railroads was fueling the ascent of iron production. In its major ore, a reddish-brown bauxite, aluminum atoms combine with uncommon tenacity to oxygen atoms as the mineral alumina (aluminum oxide). This is the basic stuff of sapphire crystals. (Alumina also is better than common silica-based clay for making tough and refractory ceramic bricks for industrial kilns.) Without electricity, yanking the metal from the aluminum ore requires the most reactive substances possible, reactants so voraciously hungry for oxygen atoms that they would rip them away from their aluminum partners. The first reactants found that could do this were the metals potassium and sodium. The late chemist and writer Primo Levi, a Jewish holocaust survivor whom the Nazi's forced to work in a synthetic rubber plant at Auschwitz, described sodium this way in his book *The Periodic Table:*

> Sodium is a degenerate metal: it is indeed a metal only in the chemical significance of the word, certainly not in that of everyday language. It is neither rigid nor elastic; rather it is soft like wax; it is not shiny or, better, it is shiny only if preserved with maniacal care, since otherwise it reacts in a few instants with air, covering itself with an ugly rough rind: with even greater rapidity it reacts with water, in which it floats (a metal that floats!), dancing frenetically and developing hydrogen.

Sodium and potassium are so reactive that in nature they always are bound with other elements into salts and minerals and never found in their pure states. Getting reasonable amounts of sodium and potassium turned out to be so difficult, expensive, and dangerous that the aluminum produced with these reactants could not compete in the marketplace.

Few things fuel innovation more than the rewards of fame and fortune. In 1886 two young entrepreneurs—Paul L. T. Heroult in France and Martin Hall in the United States—each working in home laboratories, hit upon the solution.

The key for both of them was electricity, which Thomas Edison, the great inventor and master of informed trial-and-error, was then pushing forward to help power the growing population of electric lightbulbs. Electricity, like chemistry and fire, can transform the raw materials of the world into different and new forms. Earlier in the nineteenth century chemists had found that electricity passing between the electrodes of a battery immersed in baths of chemical solutions often elicited profound effects. Beautiful mineral and metallic crystals often would build up on one electrode. Gases would often evolve from another.

Heroult and Hall—who was moved to go after the aluminum trophy by a quip from his professor about the riches in store for whoever won aluminum from rock—hit upon the right electrochemical process within weeks of each other. First they melted cryolite, a naturally occurring aluminum salt (a kind of mineral) from Greenland, in a ceramic pot lined with carbon, which served as one electrode of what amounted to a giant battery. The other electrode was in the form of a carbon rod lowered into the vat. Heroult made these rods in a high-temperature firing of powdered coke (roasted coal) bound with tar.

In molten cryolite the alumina ore dissolved like ice in warm water, yielding an alumina-rich liquid. The alumina's oxygen and aluminum atoms could be yanked apart by a strong electrical current passing between the electrodes. The oxygen combined with the electrode's carbon in a combustion reaction. Meanwhile, the aluminum atoms coalesced in the cryolite and sank to the bottom, where the metal could be siphoned off. A major difference in Hall's technique was that he used several smaller anodes rather than the one large anode that Heroult used. Both filed patents in 1886.

In 1888 Hall established the Pittsburgh Reduction Company, the progenitor of Alcoa (Aluminum Company of America), in Kensington, Pennsylvania, and began producing aluminum at a rate of 5 pounds each day. And in Europe, the Swiss Metallurgical Company in Neuhausen was the first firm to use Heroult's electrolytic cells. Within a few years a drastic drop in aluminum prices—from twelve to two dollars per pound—made widespread experimentation with the metal economically practicable. Since then, the aluminum production process has changed largely in detail and scale.

Nevertheless, steel remained the king of metals. The journalistic hype of aluminum as "silver from clay" didn't quite catch on. Ironically, steelmakers found that small additions of aluminum reduced the amount of dissolved air in the molten steel, which in turn reduced flaws in the solidified metal. The use of aluminum in steelmaking rescued aluminum production from complete unprofitability.[18]

That dependency on steel didn't last long, however. Armies experimented with aluminum as a lightweight material for mess kits. Musical instrument makers dabbled with it to make horns of various kinds, although the metal's softness doomed it for that purpose. Aluminum found its way into artificial limbs, bathtubs, jewelry, tableware, and many other products, none requiring very high volumes of material. Pure aluminum was not especially strong, and that relegated it to uses in which it would not need to withstand much stress or abuse.

Aluminum alone might not have been the next best thing since Bessemer steel, but metalworkers knew there was a world of potentially more useful aluminum alloys just waiting to be discovered. Throw in, say, a little copper or nickel, and then cool it slowly or maybe quickly, and see what you get.

It was Alfred Wilm, then head of the Metallurgical Department at the German Center for Scientific Research near Berlin, who hit paydirt. In 1902 the German Waffen and Munitionsfabriken, a consortium of German war munitions companies, asked Wilm to develop a lighter-weight alloy that might replace the heavier brass alloys so easily drawn into tubes and therefore so convenient for making bullet and munitions cartridges. Wilm looked to aluminum as a candidate. He knew that aluminum itself was not fit for the job but that an aluminum alloy might work. It didn't take long for him to find partial success with an aluminum-copper alloy. It had some of the "drawability" that the Germany munitions companies wanted, but Wilm didn't think it was strong enough.

In 1906 he embarked on a fateful set of experiments. To the aluminum and copper he added some magnesium and manganese, which had been discovered empirically decades earlier to strengthen steel. When he tested the resulting alloy for hardness by pressing a steel ball into it, the resulting impression was deep enough that Wilm assumed his experimental aluminum alloy was too soft to be useful. Two days later, however, he and an assistant checked the result by redoing the hardness test. To their amazement, the impression was much shallower. Somehow the alloy had gotten harder just by sitting there. The hardening trend continued for two more days. In the end the aluminum alloy was three times harder (and thereby probably stronger) than after the original test.

Wilm had discovered "age hardening," which metallurgists later attributed to the development of tiny mineral particles—visible in metallographic preparations—that effectively dammed up motion amongst the tiny grains that made up the metal. The discovery showed that metal is not some unchanging stuff once it cools into a solid. Like a person, it can change with time. And that meant that the way you nurture metal can make it better or worse. (It turns out that the metallurgists who made an airplane engine block for the Wright brothers, the first men to fly, had unwittingly pulled off age hardening in an aluminum alloy several years earlier.)

Three years later, Wilm's alloy, called Duralumin, went into commercial production. Of the nearly 13 tons produced in 1910, 10 went to the Vickers Company in England, which used it to build a dirigible called the *Mayfly*. Unfortunately, as workers moved the massive ship from its construction hangar in preparation for its first flight, it broke in two. Even though the disaster was the result of mishandling, the British viewed this German Duralumin as a new material to avoid.

Nonetheless, by 1914 the German Navy had specified Duralumin as the material for zeppelin airframes. In flying machines lightweight materials that are strong enough to serve as structural components are of paramount importance. Ninety-seven zeppelins were built, the larger ones requiring 9 tons of Duralumin. Aluminum as a material and aerospace as a technology joined in a common ascent through the ensuing decades.

The sporadic inventions and discoveries of the nineteenth century—new polymers like celluloid and Bakelite and new metals like aluminum and Duralumin—were concrete demonstrations that there were many stones yet to be unturned—or fortuitously chipped—in the realm of stuff. In the twentieth century a new piston—sociological innovation—joined curiosity, accident, and economic incentive in speeding the engine of exploration.

The era in which individuals like Goodyear, Hyatt, Wilm, and Baekeland could spawn industries off of a single material discovery was giving way to a higher-powered, more systematic approach that relied more on the collective brain of institutional science. As Bakelite diffused into people's homes and aluminum showed its aeronautical mettle overhead, the race for still better and newer materials was accelerating in new industrial laboratories that sought to hasten the transition from theoretical insights to marketable commodities.

Edison set the precedent in 1876 when he opened his laboratory in Menlo Park, New Jersey. He told the world that its output would be "a minor invention every ten days and a big thing every six months or so."[19] The German dye and chemical industry had begun systematic chemical research in the 1880s, resulting in an economic coup that shifted the center of chemical innovation from England to Germany. General Electric set up its first industrial laboratory in 1900 initially to improve the company's lightbulb technology. Two years later E. I. du Pont de Nemours and Company followed suit. Andrew Carnegie, the steel mogul and railroad baron, did the same and wrote in his autobiography that "nine tenths of all the uncertainties of pig-iron making were dispelled under the burning sun of chemical knowledge." Eastman Kodak, American Telephone and Telegraph, Corning Glass Works, and many, many others soon joined the trend.

This sociological sea change in the mode of materials discovery showed up most dramatically in the proliferation of new polymers. In a laboratory at B. F. Goodrich in 1926, a chemist made polyvinylchloride, from which he made adhesives and sheets. Four years later the German chemical giant I. G. Farben introduced polystyrene, which could even be made into extremely lightweight foams. In 1935 another firm, Rohm and Haas, rolled out polymethyl methacrylate (first discovered over sixty years earlier) as Plexiglas, which was at once transparent, strong, and tough enough to substitute for glass over the cockpits of military airplanes. Two years later DuPont rivaled Plexiglas with its own Lucite.

Within the next ten years came epoxy resins, Teflon (an accidental discovery whose nonstick properties are due to the four fluorine atoms on each of its molecular building blocks), polyethylenes (which are ideal for forming containers like milk jugs), polyethylene terephthalate (strong and impervious enough to contain pressurized liquids like soda), and others. The number and rate of new polymer discoveries has only increased since then.

Of all of the polymer developments, none was more emblematic than the invention of nylon at DuPont's Purity Hall in Wilmington, Delaware.[20] The process started in 1928, when the company lured a brilliant, thirty-two-year-old chemist, Wallace Carothers, from his university bastion at Harvard. His mission was to unravel the chemical details behind the formation of polymers, whose general molecular structure at the time was under intense debate. Some chemists thought they were made of small molecules bunched together like minuscule clumps of dust; others held that polymers were essentially enormous versions of the small molecules that chemists had gotten used to over the previous hundred years. Carothers approached polymers using the most current techniques and concepts of science, including the new ideas of quantum mechanics that were beginning to permeate the ranks of chemists.

Within three years of Carothers's arrival, his group scored a hit: they developed neoprene, one of the first commercial synthetic rubbers. Far more resistant to chemicals than vulcanized natural rubber, it was well suited for use under the hoods of automobiles. After that, the lab focused on the quest for artificial textile fibers. Rayon, pretty much a fiber version of cellulose from plants, already had reached commercial production in the 1920s. Carothers and his colleagues wanted to untether themselves from the restrictions of naturally occurring feedstocks such as cotton-derived cellulose. They wanted to start from chemical scratch.

For a model they turned to silk, the strong, resilient, and supple fiber produced by spiders and insects like the silkworm. Silk fibers are made up of protein molecules, which are natural polymers themselves composed of amino acid molecules linked into chains. The chemical linkage between amino acids is known as an amide bond, and that was one of the chemical domains that Carothers's team decided to reconnoiter. The proteins that cellular chemistry builds into proteins could also be called polyamides. So the DuPont team was out to make artificial polyamides.

Each formulation would consist of one or more types of small organic molecules, somewhat like the amino acids of proteins, that would be linked into polymers under various reaction conditions. One day in May 1934, one of Carothers's assistants put a stirring rod into a beaker containing the syrupy result of a reaction. When he pulled the rod out, a long string of the liquid followed up like the tail of a kite and quickly cured into a silklike fiber. This became known by polymer chemists as "the rope trick." That rope trick led to the first nylon fibers. By 1937 DuPont researchers spun the first nylon yarn and wove the first experimental stockings. That same year Carothers, who had been suffering from increasingly severe depression, committed suicide with a dose of cyanide in a Philadelphia hotel room. It was a mere three weeks after DuPont had filed for the basic patents on nylon. In 1938 nylon went into production.

Nylon was instantly popular. In 1939, when the first nylon stocking went on sale, DuPont promoted its new silk substitute with fervor, resorting even to 30-foot-tall mannequin legs touting the new stuff. World War II deflected this zealous marketing campaign for a while. When Asian supplies of silk were cut off, the U.S. government turned to the new synthetic silk—nylon—for the tough material it needed for parachutes, tire cords, flak vests, and rope. For several years all nylon that DuPont produced went into the war effort; none was sold in commercial markets.

After the war the material's combination of silklike toughness and elasticity as well as its better resistance to chemicals and electricity made it attractive for all kinds of things. As a fiber it made it into everything from shirts to climbing ropes. But it also could be molded into hundreds of items, including combs, machine gears, zippers, and electrical insulators. Modifications of its chemical constituents and the way it is processed have continued the diversification of its roles. Nylon remains a household name and continues to be part of DuPont's bread and butter.

The mushrooming of industrial materials laboratories was part of a larger trend that had begun to accelerate in the first decades of the twentieth century: the interdisciplinary quest for the key to the secret inner life of materials. As physicists dissected the anatomy of atoms, metallurgists, ceramic experts, chemists, and all kind of engineers and scientists delved in ever more detail into the relationship between a material's many-leveled structure and its properties. Why exactly do ceramic materials break so readily but metals only dent? Why does steel resist rusting when you put enough chromium into it? What is going on inside of a piece of metal when it undergoes age hardening? What makes rubber so elastic? Can a carefully constructed crystalline material function like an entire device, say, like the vacuum tubes of early electronic technology?

Thousands of scientists and engineers in hundreds of different labs began to look into questions like these. At best, they were peripherally aware of one another's pursuits. Their common thread was that they were routinely validating the importance of Réaumur's prescient hierarchical portrayal of materials in the eighteenth century, his intuition that a material's essence depends upon how its ingredients assemble from the atomic level up to the macroscopic level. They were also validating and extending Sorby's call to observe and reveal that hierarchy in as much detail as possible. Like a person's personality, every material's collection of traits and behaviors comes from what it is made of and how those ingredients were nurtured into their present form. Not only did researchers in the first half of the twentieth century know this intellectually, but they were also building the analytical tools and concepts that finally were about to let them uncover the details in practice.

THREE

the SECRET ARCHITECTURE OF STUFF

It takes an uncommon technical breadth to traverse matter's vast inner worlds of atoms and molecules and its equally expansive outer world of diverse substances and objects. It requires an intricate blend of chemistry, physics, mathematics, computer science, engineering, and whatever other field of knowledge or skill happens to help. The ultimate task, after all, is to understand how a common palette of about ninety types of atoms—the periodic table of the chemical elements—can harbor earth, air, fire, water, flint, obsidian, malachite, quartz, copper, bronze, iron, stainless steel, clay, pottery, porcelain, glass, Portland cement, limestone, rubber, cellulose, Bakelite, nylon, horn, olive oil, petroleum, spandex, cotton, silk, electronics-grade silicon, gunpowder, skin, bone, and countless other materials.

The reductionist approach of trying to understand materials in terms of atoms and molecules alone is only narrowly instructive. Atoms and molecules are indeed the most basic ingredients of materials. But it's important not to miss the

FIGURE 3

The Periodic Table of the Elements

1	2	3	4	5	6	7	8	9	10	11	12	13	14	15	16	17	18
IA																	VIIA
1 H	IIA											IIIA	IVA	VA	VIA	VIIA	2 He
3 Li	4 Be											5 B	6 C	7 N	8 O	9 F	10 Ne
11 Na	12 Mg	IIIB	IVB	VB	VIB	VIIB	——	VIII	——	IB	IIB	13 Al	14 Si	15 P	16 S	17 Cl	18 Ar
19 K	20 Ca	21 Sc	22 Ti	23 V	24 Cr	25 Mn	26 Fe	27 Co	28 Ni	29 Cu	30 Zn	31 Ga	32 Ge	33 As	34 Se	35 Br	36 Kr
37 Rb	38 Sr	39 Y	40 Zr	41 Nb	42 Mo	43 Tc	44 Ru	45 Rh	46 Pd	47 Ag	48 Cd	49 In	50 Sn	51 Sb	52 Te	53 I	54 Xe
55 Cs	56 Ba		72 Hf	73 Ta	74 W	75 Re	76 Os	77 Ir	78 Pt	79 Au	80 Hg	81 Tl	82 Pb	83 Bi	84 Po	85 At	86 Rn
87 Fr	88 Ra																

57 La	58 Ce	59 Pr	60 Nd	61 Pm	62 Sm	63 Eu	64 Gd	65 Tb	66 Dy	67 Ho	68 Er	69 Tm	70 Yb	71 Lu
89 Ac	90 Th	91 Pa	92 U	93 Np	94 Pu	95 Am	96 Cm	97 Bk	98 Cf	99 Es	100 Fm	101 Md	102 No	103 Lw

FIGURE 4

The Elements in Ceramic Materials

1	2	3	4	5	6	7	8	9	10	11	12	13	14	15	16	17	18
IA																	VIIA
	IIA											IIIA	IVA	VA	VIA	VIIA	
3 Li	4 Be											5 B	6 C	7 N	8 O		
11 Na	12 Mg	IIIB	IVB	VB	VIB	VIIB	——	VIII	——	IB	IIB	13 Al	14 Si	14 Si	15 P		
19 K	20 Ca	21 Sc	22 Ti	23 V	24 Cr	25 Mn	26 Fe	27 Co	28 Ni	29 Cu	30 Zn	31 Ga	32 Ge				
37 Rb	38 Sr	39 Y	40 Zr	41 Nb	42 Mo	43 Tc	44 Ru	45 Rh	46 Pd	47 Ag	48 Cd	49 In	50 Sn	51 Sb			
55 Cs	56 Ba		72 Hf	73 Ta	74 W	75 Re	76 Os	77 Ir	78 Pt	79 Au	80 Hg	81 Tl	82 Pb	83 Bi			
87 Fr	88 Ra																

57 La	58 Ce	59 Pr	60 Nd	61 Pm	62 Sm	63 Eu	64 Gd	65 Tb	66 Dy	67 Ho	68 Er	69 Tm	70 Yb	71 Lu
89 Ac	90 Th	91 Pa	92 U	93 Np	94 Pu	95 Am	96 Cm	97 Bk	98 Cf	99 Es	100 Fm	101 Md	102 No	103 Lw

---

The Atomic Stuff of Stuff. The periodic table of the chemical elements represents the ultimate pantry of ingredients that make up all the materials there ever can be on earth. At the top is the entire table, showing all of the elements except for the last half-dozen man-made elements that would go

FIGURE 5

The Metallic Elements

1	2	3	4	5	6	7	8	9	10	11	12	13	14	15	16	17	18
IA																	VIIA
	IIA											IIIA	IVA	VA	VIA	VIIA	
3 Li	4 Be											5 B					
11 Na	12 Mg	IIIB	IVB	VB	VIB	VIIB	——	VIII	——	IB	IIB	13 Al					
19 K	20 Ca	21 Sc	22 Ti	23 V	24 Cr	25 Mn	26 Fe	27 Co	28 Ni	29 Cu	30 Zn	31 Ga					
37 Rb	38 Sr	39 Y	40 Zr	41 Nb	42 Mo	43 Tc	44 Ru	45 Rh	46 Pd	47 Ag	48 Cd	49 In	50 Sn	51 Sb			
55 Cs	56 Ba		72 Hf	73 Ta	74 W	75 Re	76 Os	77 Ir	78 Pt	79 Au	80 Hg	81 Tl	82 Pb	83 Bi			
87 Fr	88 Ra																

57 La	58 Ce	59 Pr	60 Nd	61 Pm	62 Sm	63 Eu	64 Gd	65 Tb	66 Dy	67 Ho	68 Er	69 Tm	70 Yb	71 Lu
89 Ac	90 Th	91 Pa	92 U	93 Np	94 Pu	95 Am	96 Cm	97 Bk	98 Cf	99 Es	100 Fm	101 Md	102 No	103 Lw

FIGURE 6

The Elements in Polymeric Materials

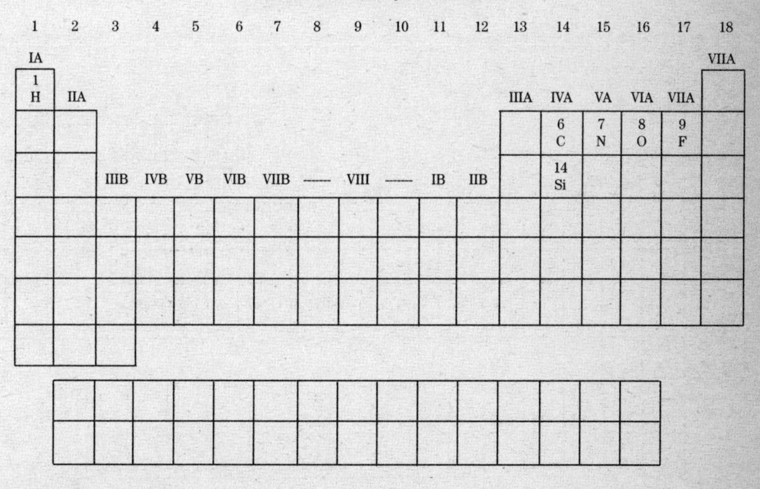

after element 103. The following five tables indicate which elements build into ceramics, metals, polymers, semiconductors, and elemental superconductors, respectively.

REPRINTED WITH PERMISSION FROM THE *CRC MATERIALS SCIENCE AND ENGINEERING HANDBOOK*, 2D ED. (BOCA RATON, FLORIDA: CRC PRESS, 1994).

FIGURE 7

The Elements in Semiconducting Materials

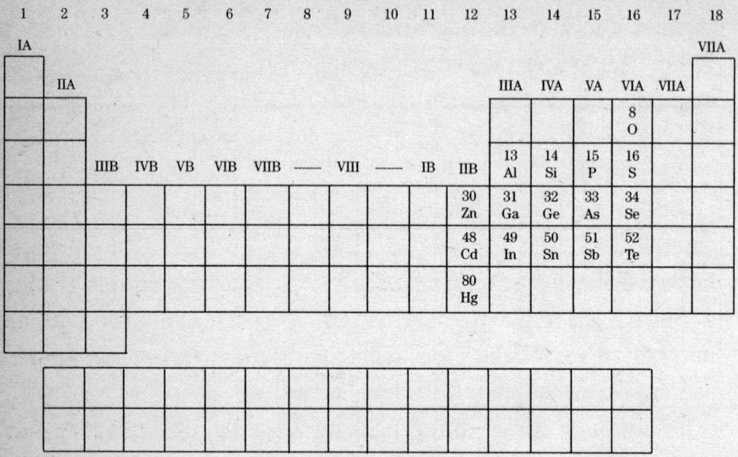

1	2	3	4	5	6	7	8	9	10	11	12	13	14	15	16	17	18
IA	IIA	IIIB	IVB	VB	VIB	VIIB	——	VIII	——	IB	IIB	IIIA	IVA	VA	VIA	VIIA	VIIA
															8 O		
												13 Al	14 Si	15 P	16 S		
											30 Zn	31 Ga	32 Ge	33 As	34 Se		
											48 Cd	49 In	50 Sn	51 Sb	52 Te		
											80 Hg						

FIGURE 8

The Superconducting Elements

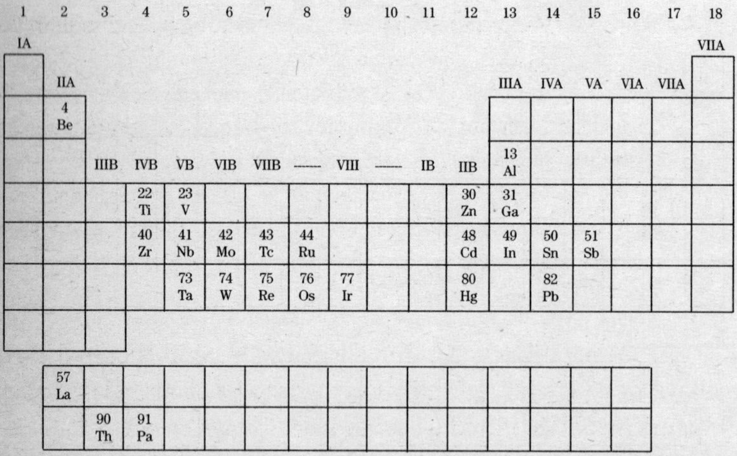

1	2	3	4	5	6	7	8	9	10	11	12	13	14	15	16	17	18
IA	IIA	IIIB	IVB	VB	VIB	VIIB	——	VIII	——	IB	IIB	IIIA	IVA	VA	VIA	VIIA	VIIA
	4 Be																
												13 Al					
			22 Ti	23 V							30 Zn	31 Ga					
			40 Zr	41 Nb	42 Mo	43 Tc	44 Ru				48 Cd	49 In	50 Sn	51 Sb			
				73 Ta	74 W	75 Re	76 Os	77 Ir			80 Hg		82 Pb				

57 La														
	90 Th	91 Pa												

forest for the trees. Just as a sociologist seeks to understand the dynamics of human interaction in lesser or greater collectivities under a variety of conditions, so the chemist tries to grasp the ways in which the elements of matter *interact*—aggregating, segregating, rearranging, mingling, and repelling to emerge as multitiered structures, each with its own set of material traits.

The hierarchical structure of a material is the result of the interaction of its ingredients. That is why a little more or less heat, a different proportion of alloying elements, a finer grade of a pulverized ingredient, this or that contaminant, and any number of variations on a standard process can result in what seems like an entirely different material.

Because these differences in process can produce such wide differences in materials, achieving the desired result is a lot like cooking. This is how a couple of modern aluminum engineers put it:

> When you mix pumpkin, spices, sugar, salt, eggs, and milk in the proper quantities, you can make a pumpkin pie filling. By adding flour and adjusting the proportions, you can make pumpkin bread. Substituting shortening for the pumpkin and molasses for the milk will yield ginger cookies. Each adjustment of the recipe results in a different dessert. Just as adding enough flour can turn pie filling into bread, adding enough chromium to steel makes it stainless steel.[1]

The metaphor holds for all of the several hundred steel alloys in use today. It holds for the tens of thousands of other materials that go into the human-made things on earth.

So a process yields a material with a distinctive hierarchical structure that determines the material's properties and how it will perform in its various applications. That is the truism that Henry Sorby helped raise to a paradigm of metallurgical research and practice. In his day a microscope was sufficient for a metallurgist to discern qualitative differences within a cut, polished, and etched metal surface. This was a level of structure above the chemical one. A chemical analysis could identify the kinds of atoms present in a piece of metal and the relative proportions of those atoms, but not how those atoms came together into specific structures. Metallography helped to uncover such structures and patterns of material behavior.

Over time a metallographer learns to recognize material meaning in the mineralogical and metallurgical montages that he sees through his microscope. A montage that looks like an aerial view of snow-covered mountains comes from gray cast iron, the dark outlines of "mountain" contours coming from graphite flakes in the metal's microstructure. Another montage has dark, needlelike outlines crisscrossing through a white background. That pattern comes from a hard steel whose iron-based crystals have assumed two distinct geometric arrangements known as austenite and martensite, named after the researchers who first identified them. The martensite constitutes the crisscrossing needles in a background of austenite.

Metallographers eventually learned to correlate these montages with certain metallurgical formulas, alloying ingredients, and processing practices: that is, to the various ways you can cook steel. By the end of the nineteenth century, metallography, chemical analysis, and other techniques for looking under a material's skin were becoming standard tools for guiding industrial metal producers toward consistently good results.

It was a growing assumption after the turn of the century that advances in materials would no longer come as easily as they had in the past. The brazen empiricism that guided the efforts of pioneers like Henry Bessemer and John Wesley Hyatt still played a role, but the heyday of the lone, hit-or-miss tinkerer was over. Advances grew increasingly out of precise quantitative observation, measurements, and the kind of theorizing that physicists and chemists had been honing over the past century.

Despite the improvements in research methodology, the limitations of materials technology remained all too apparent. In the first years of the twentieth century, for example, the annual average accident rate on the U.S. railroad system was about four thousand, resulting in approximately thirteen thousand deaths and injuries per year. The blame for most of the carnage went to metal gone bad: broken rails, wheels, flanges, and axles pushed to failure by a combination of excessive loads, inadequate maintenance, and inferior iron and steel.[2]

But the toll from inadvertent death paled next to the purposeful slaughter of World War I, which became a merciless laboratory for materials research in which the potentially beneficent tools of modern science—advanced machinery, mass production, precision manufacturing—were converted into instruments of death. Even chemistry diverted some of its attention from the boon of pharmaceuticals toward fostering a plague of chemical weapons in what some historians have called "the chemists' war." "It was [also] an iron and steel war, with comparative advantage in weapons technology determined by relative superiority among belligerents in steel-ingot forging, foundry, and metalworking technologies," notes one metallurgist who later worked on the unprecedented materials problems related to nuclear energy.[3] Metallurgists were hidden heroes—or, perhaps, in the eyes of the victims, villains—in World War I, just as they were for the Hittite armies and Samurai warriors of centuries past.

Tools like metallography—which can help reveal the structural links between a material's levels, from the invisible atomic base to the visible superstructure—opened only some of the many windows on material anatomy. A metallographer couldn't tell from his microscopic observations exactly how the atoms in each region or grain of a metallographic preparation were arranged. Were the atoms tightly packed like perfectly stacked oranges, or were they arrayed more randomly, as though there were lemons and potatoes in the stack? Or were crystalline utopias mixed with anarchic dystopias within an apparently unitary metallographic spot?

Deep internal differences like these, it turns out, have everything to do with a material's mechanical properties: its strength, its resistance to fracture, its susceptibility to corrosion or deformation. At the turn of the century, the key to enhancing understanding of and control over a metal's internal structure and thereby over its technological abilities lay in seeing structures finer than those discernible in Sorby's metallography.

X rays, first observed in 1895, provided one very powerful means. The study of crystal structure first developed mathematically in the eighteenth and nineteenth centuries. It exploded into experimental significance in 1912 and 1913, when the father-son team of William Henry Bragg and William Lawrence Bragg developed a technique called X-ray crystallography, which uses X rays as a precise tool for measuring the crystal's structural regularities on atomic scales. Here's how it works: as X rays pass into a crystal, they diffract (a kind of atomic-level reflection) from the crystal's regular arrangement of atomic columns and planes, which act like mirrors to the incoming X rays. When the Braggs let these diffracting X rays expose photographic film, the result was a diffraction pattern that disclosed the crystal's regularities or lack thereof. From the photographic pattern of rings and spots, arranged more or less concentrically, crystallographers can calculate things like distances between adjacent atoms (or molecules) and atomic (and molecular) planes.

In principle the technique provided a zoom lens on the individual crystal grains visible to metallographers on their optical microscopes. It enabled the metallurgist to link empirically the properties of a piece of steel or brass all the way down to its atomic idiosyncrasies. At the same time, X-ray crystallography underscored the reality that real materials seldom exist in perfectly crystalline states. Most often they are conglomerations of many crystals, just like most stones you might pick up from the ground. Or, like glass and polymers and liquids and gases, they have little or no internal structural order. Send X rays into a disordered or amorphous material and they diffract every which way, leaving on the film a cloudlike image providing a paucity of detailed structural information.

As the Braggs gave the world a window on the atomic level of material structure with their X-ray crystallography, Alfred Wilm's discovery of age hardening was opening another window onto the structure of materials. Three researchers at the U.S. National Bureau of Standards (now called the National Institute of Standards and Technology) actually teased apart the microstructural details underlying Wilm's observations. The NBS researchers discovered that the solubilities of the alloying elements in Duralumin were much higher in solid hot aluminum (at the 525°C or so that Wilm had used in his work) than at room temperature. Once the metal cooled to room temperature from the heat treatment, scientists observed some new formations within the sea of aluminum: tiny crystal islands with chemical compositions such as $CuAl_2$ and Mg_2Si; and tiny zinc particles. The size and distribution of these islands mattered. Hard alloys had lots of small particulates. But a smaller number of larger particulates, or too many of the smaller particulates, led to weaker metal that fractured more easily. The NBS researchers had found a key to what makes aluminum and many other metals good or bad.

As usual in science, solutions become the problems. An aluminum alloy containing a certain distribution of mineral and various metallic particulates might be harder than one with some other distribution. But how in fact does structural detail actually harden the metal? It is one thing to observe structures that correlate with material properties like hardness. It is another to understand the mechanism by which the structure yields the property in question. If you understand the mechanism of, say, age hardening, then you have a better chance of figuring out how to make harder or stronger metals.

By 1930 metallurgists and physicists had postulated several kinds of large and small imperfections within and between a metal's grains. These structural imperfections turned out to be at least as important to a material's properties as its chemical composition and the molecular structures within its crystal grains. As Cyril Smith once wryly put it, "Everything really interesting about materials resides in the hierarchy of their imperfections."[4]

Like biologists spotting new species in a previously unexplored ecosystem, an army of scientists, equipped with ever better analytical instrumentation (including electron microscopes beginning in the 1930s) were happening onto new kinds of defects and microstructures and determining the processes by which these formed. Among the defects is the absence of an atom from its normal place in a crystal (a vacancy); another is a single contaminant atom taking the place of an expected atom in a particular crystal. It is like finding one tangerine or grapefruit in a stack of naval oranges.

Contaminants also tend to diffuse more rapidly along larger defects such as the boundaries between crystal grains and dislocations within atomic planes of the grains. You can think of boundaries as defects since the most perfect piece of metal would be one huge monolithic crystal with no internal boundaries. Like dirt between two pieces of wood glued together, atomic riffraff between metal grains weakens the intergrain bonds that hold a piece of metal together. An example of such dislocation is a line defect, which provides internal slack for atomic planes to slide by one another. Line defects help determine a metal's ductility, formability, and vulnerability to various stresses and strains.

Another type of defect is known as a screw dislocation,

which results from atoms linking to one another in an upwardly helical sequence that allows growth to occur more continuously than if one atomic plane had to be fully completed before the next could start. Screw dislocations also govern the strength and chemical performance of materials. Atoms in the helical structure can be more vulnerable to the kinds of chemical attacks that lead to corrosion, embrittlement, and other nasty changes. Such imperfections also can be good, however, since it is the screw dislocations, rather than the metal crystals hosting them, that suffer the violence. Such balancing acts are typical of the quest to make good materials.

Such defects and processes have become the meat and potatoes of research into the mysteries of metallic qualities such as hardness, softness, ductility, and brittleness. Growing understanding of them has opened pathways to making new metallic alloys and ceramic materials that are harder or more deformable or tougher than previously known materials.

Although studies and experiments by metallurgists spurred many of the big advances in materials research, cross-pollination of insights among chemists, physicists, and other scientists was also generating new sets of rules for revealing and exploiting the similarities and differences among materials. Thanks to this growth of multifaceted knowledge about how specific defects, phases (different states of matter, including different crystallographic forms in a single solid sample), and other microstructural features affect a material's properties, a more rational approach to alloy and material development was becoming possible.

By World War II researchers were routinely taking Henry Sorby's microstructure-to-macrofunction paradigm to atomic levels of detail. Metallography, X-ray diffraction, electron microscopy, and other analytical techniques often provided the data. There were many tools for making sense of the data: dislocation theories; thermodynamic models that described how different solid phases of different compositions and crystal forms end up in materials; and ideas about precipitation hardening and related concepts. Metallurgy thus became connected to other disciplines investigating solid materials such as minerals and ceramics, the mechanical properties of which were affected by the same things.

The convergence of ideas and research techniques gradually brought members of these related fields together under one umbrella that came to be known as solid-state physics. In the 1930s and 1940s this burgeoning discipline was the intersection where material arenas as seemingly disparate as polymers, metals, and ceramics converged.

A major by-product of this interdisciplinary ferment was the proliferation of polymers. The celluloid plastics were already nearly sixty years old. Bakelite had been around for a few decades. Each year industry researchers were unveiling new polymers with new properties and technological possibilities. The Age of Plastics had begun.

Many of the polymer players were sophisticated theoretical chemists, but the development of the materials still advanced largely by dogged empirical work. Try this or that reaction. Try this or that temperature or pressure. Throw in a little metal and see if it catalyzes different reactions. Moreover, polymeric microstructure is rarely crystalline. It is more like cooked spaghetti. It is more complicated and harder to unveil than a standard metallic or mineral microstructure. Shine X rays through most synthetic polymers and the picture you get on the film is a diffuse cloud. That is just what you might expect from a jumble of molecules resembling a pasta dish.

Not only were researchers clueless as to how polymer molecules clump together to form a material, but until the mid-1930s, most chemists didn't even acknowledge that polymeric materials were made of huge molecules. The late Herman Mark—an Austrian refugee who spent his professional career at

polymers

the Brooklyn Polytechnic Institute, where he rose to become one the fathers of polymer science—put the situation in the late 1920s this way: "There were fiber chemists; cellulose chemists, a Society for Cellulose Research, books on cellulose, journals on cellulose. The same thing for silk, the same for wool, for rubber, and starch. So there were five different disciplines, strongly represented by people, by literature, books, societies, and so on."[5]

The modern field of polymers supplied a common thread to this disjointed quest when the German chemist Herman Staudinger began floating what was then a laughable idea to most other chemists:[6] he argued that natural materials such as silk, rubber, and starch, as well as lab-made materials such as Bakelite rayon and celluloid, were all composed of huge chainlike molecules. He was proposing the existence of enormous molecules hundreds or thousands of times larger than the smaller, more compact molecules with which chemists had become familiar during the past century.

Herman Mark likened Staudinger to a biologist asking his colleagues to believe there were elephants 400 feet long and 200 feet high actually roaming the earth.[7] A more popular idea for explaining the properties of polymers (such as their higher viscosity compared to liquids like water or benzene) was that smaller molecules congregated into larger structures without actually forming strong chemical bonds. The resulting so-called colloidal structures were more like a bunch of people squeezing together as in a Rugby scrum or a New Age hug.

By the mid-1930s, however, most chemists had come around to Staudinger's way of thinking. The idea of long molecules made of different smaller building blocks that link like beads of a necklace was a powerful unifying principle. Thanks to Staudinger's daring idea, what Mark had noticed as five separate disciplines in the 1920s now evolved into subdivisions of a single, increasingly fertile field. Wallace Carothers, an industrial polymer scientist at DuPont, adopted and applied Staudinger's theory with momentous practical results: he invented neoprene (a chemically resistant synthetic rubber) and nylon, two immensely useful and profitable new materials.

It was just the beginning for polymers, of course. If you could take any number of building blocks and then modify and link those in any number of ways, you had in your hands a fountain of new materials that could change the industrial and social landscape. One of the most famous cinematic sound bites of all time is testimony to this versatility and impact of polymer science. In the 1967 film *The Graduate*, Dustin Hoffman, playing a recent college graduate who is alienated and adrift in his parents' prosperous, materialistic suburb, receives one word of career advice from a friend of his father's: "Plastics." That memorable utterance stood as both a tribute to plastic's pervasive impact on society and as a sardonic reminder of its dubious aesthetic and spiritual legacy for the young radicals of the Sixties.

While Staudinger was advancing his pathbreaking hypothesis in the mid-1920s, an all-encompassing conceptual unification known as quantum mechanics was under way. It would join under one theoretical umbrella metals, ceramics, polymers, and every other material that ever was, is, or will be.

Quantum mechanics was an irksome answer to the various troubles with the nineteenth-century belief, stemming from the ancient Greek atomists Democritus and Leucippus, that atoms were supposed to be hard, impenetrable spheres—in effect, microscopic billiard balls. But two discoveries suggested that atoms did in fact have some kind of differentiated internal structure: 1) all atoms have electrons (which are far less massive than the smallest atom), and 2) there is radioactive emission of particles from atoms. If atoms had parts, they could not be teeny spherical monoliths.

There were other reasons to think that atoms had some internal structure. For decades, spectroscope-peering scientists were delighted and mystified by the characteristic set of colors—a spectrum of light—that different elements emitted and absorbed. Every time you light a candle, you witness the emission of colored light from the excited electrons in the atoms and molecules of the candle's wick and wax. The nineteenth-century scientists knew that the colors somehow could be due to electrical charges in atoms vibrating at different frequencies. These oscillations would produce electromagnetic radiation that could show up as the element's different spectroscopic colors.

But the idea of atoms with moving parts was hard to reconcile with other phenomena that made more sense if the atoms were in fact like little billiard balls. The century's most brilliant and powerful physicists were publishing, speaking, and pushing into the scientific canon the "kinetic theory of gases." Billiard-ball-like atoms in motion, they asserted, explained all kinds of things, like the pressure of gases at different temperatures, the way heat travels through different materials, and why crystals assume the shapes they do. More complicated atoms with moving parts would muddy these apparently clarified waters.

Was there some picture of an atom that could combine the best of both worlds, accounting for the moving parts implied by the spectroscopic observations yet also preserving the more monolithic, billiard-ball character essential to the prevailing physical theories?

In the fall of 1900 at the University of Berlin, the physicist Max Planck further confounded orthodoxy with his finding that wavelike, spread-out forms of energy like light and heat seem to be made of discrete packages, or quanta, just as matter does. This insight marked the birth of quantum mechanics, which has kept theorists ever since scrunching their faces trying to explain how matter and energy can behave simultaneously like discrete particles and spread-out waves.

Quantum mechanics pulverized the cornerstones of classical physics, which had doggedly insisted on a more mechanical model of the universe. In the traditional model, energy ought to be able to assume any value, not just discrete ones like specific points on a line. After all, doesn't a piece of hot metal smoothly cool so that it takes on every single temperature between its hottest point and its lowest? No, said the quantum mechanicians. A piece of matter like iron only appears to cool down continuously without jumps; the discontinuities in the quantum emissions were indiscernible because a typical piece of visible matter consists of billions upon trillions upon quadrillions of individual particles, each of which cools down by its own set of quantized emissions of energy. But aggregated in the piece of iron, these countless particles create a statistical illusion of continuous cooling.

In 1913 a young Danish physicist named Niels Bohr had an idea of how these revelations might affect the portrait of atoms. What if the atom consisted of a tiny heavy nucleus with featherweight electrons orbiting around it like planets around a sun? Bohr's solar system model gave electrons, which some scientists even thought could be the basic stuff of all atoms, a definite role in atomic structure. It also nicely assimilated the earlier discovery that the vast bulk of atomic mass resides in a very small positively charged nucleus that spanned only one one-hundred-thousandth or so of the atom's diameter.

The solar system model also suggests that lines and colors seen in atomic spectra correspond to changes in an electron's orbit around the atom's nucleus. When an electron absorbs just so much energy—a specific quantum of energy—it jumps up to the next orbital. In so doing, it prevents the color of light corresponding to that absorbed energy from reaching a spectroscope, thereby leaving a black line characteristic of absorption spectra. An already excited electron emits just so much light energy—again, a specific quantum—when it relaxes to a lower orbital. In so doing, the electron emits light of a color that shows up in an emission spectrum. With Bohr's ideas, atomic structure and behavior seemed to be falling into place.

But pretty as it was, the solar system idea was vague. It revealed nothing about why atoms of different elements have such different chemical behavior. It failed to explain why electrons orbiting nuclei don't eventually lose their energy and careen into the nucleus the way a bird that stops flapping will eventually crash to the earth. It couldn't explain why copper was conductive but rubber was not or, for that matter, why any material had any particular trait.

There were other terribly vexing problems. Physicists were discovering that the pointlike electrons often behaved as though they were spread-out waves. Send them through crystals and they would produce ripply interference patterns much like those of waves passing through a gallery of poles at a pier. If electrons were like tiny billiard balls, they would have produced a much more discrete pattern. Yet if electrons were in fact more spread out, like waves, then they didn't fit neatly into the well-defined orbits that Bohr had in mind.

With light and particles each capable of having characteristics of the other, the neat and clean Newtonian picture of billiard balls in motion was banished to fantasyland. Newtonian physics still worked as a practical and convenient tool for the large objects (like *real* billiard balls) that mechanical engineers and ordinary mortals dealt with. But it seemed to have less and less to do with the microscopic domains that scientists and industrialists were also beginning to understand and control.

The quantum mechanical portrait of the atom that finally stuck reflected the maddening dualisms and ambiguities that no one could dispel from the atomic world. Add to that Werner Heisenberg's famous principle of ineluctable uncertainty about matter on these smallest subatomic scales (you cannot know simultaneously the exact location and motion of an electron, for example), and all pictures of atoms and their interactions on these quantum levels were becoming a blur. Light, long thought to consist of waves of energy, now also had to be thought of as some kind of particle. And particles, like electrons and atoms, had to be thought of also as having wavelike properties. The location of electrons around nuclei now could be expressed only as probabilities of being more likely here than there. Knowledge of atoms and hence of matter itself would forever be haunted by uncertainty.

The Austrian physicist Erwin Schrödinger brought some mathematical exactitude to this portrait with an equation that has made him famous. His equation sits in the scientific pantheon alongside Einstein's $E=mc^2$ and Newton's $F=ma$. Schrödinger's equation has become a mathematical guide for calculating wave patterns in a space or field infiltrated by entities like electrons and protons that can influence that field. The equation's raw output, which is called a wave function, is a map of the likelihood that a particle such as the lone electron of a hydrogen atom will be found in a specific region of space. The precise size, shape, and contours of this map depend on the number and locations of other electrons and atomic nuclei.

With that spin the Schrödinger equation assumes an awesome power, in principle at least. As a modern physical chemistry textbook puts it, "A wave function contains ALL THERE IS TO KNOW about the outcome of experiments that can be done on a system," which includes the making of all materials.[8] Within that capacious equation, then, is the chemical behavior of all atoms and compounds and the answer to whether any proposed chemical reaction will work or not. It can demystify all known chemical behavior and predict chemistry as yet unobserved. It can account for how copper was relatively easy to smelt nine thousand years ago, how six carbon and six hydrogen atoms form into benzene's planar circular shape, and how a contemporary scientist might be able to develop a new polymer that can convert red light into blue light for higher-capacity communications systems.

Schrödinger's equation did not put chemists and materials researchers out of business because it is, to put it mildly, a mathematical bitch. It is a computational Mount Everest. Accurately computing the wave functions (and thereby the chemical and physical behavior) of even the simplest atom—the hydrogen atom (one proton and one electron)—was impossible until the 1960s, when computer power began becoming more widely available. Even to solve Schrödinger's equation for a hydrogen molecule (two protons and two electrons in motion) requires one of today's most powerful supercomputers. For molecules with a half-dozen atoms, each with their fair share of electrons, solving the raw equation is beyond computational reach.

Yet Schrödinger's equation and its quantum mechanical framework did probe the insides of materials with an entirely novel kind of lens. Scientists became adept at developing clever approximations and shortcuts that cut the equation down to more manageable though less accurate forms with which they could extract at least qualitative answers to their questions. It helped physicists understand how the electrons in some materials can start to flow in electrical currents while not in others (see chapter 8). And the quantum mechanical framework yielded a flood of insights about how matter and energy in forms such as light, heat, and electricity can interact in almost magical ways.

Here, it seemed, was a Rosetta stone for unearthing wondrous new materials. Consider what happened just before Christmas in 1947 at Bell Telephone Laboratory in suburban Murray Hill, New Jersey.[9] There, Walter Brattain and John Bardeen watched an oscilloscope trace out the amount of electricity passing through an inelegant-looking gadget that Brattain had made. It consisted of a crystal block of germanium (peppered with atoms that could accept electrons) on which there were two tiny gold spots separated by about 4 microns (a hair is about 50 microns in diameter). (See Figure 9.)

When the physicists applied a tiny electric voltage to one gold contact, the electronic environment in the crystal changed.

FIGURE 9

The Mother of All Transistors. In late 1947 researchers at Bell Laboratories put together inelegant-looking yet revolutionary gadgets like this germanium-based point-contact transistor, which could amplify electrical signals. At the bottom of the plastic triangle are two tiny spots of gold separated by about two thousandths of an inch. When the researchers, who later became Nobelists for their work, touched these gold contacts to the underlying slab of germanium and then applied a voltage to one of the contacts, the amount of current flowing into the germanium crystal through the other contact increased. This amplification became known as the transistor effect and became a central pillar of the microelectronics revolution.

BELL LABORATORIES

A crowd of positive charges formed near the crystal's surface. The internal result was a new rapid-transit pathway for electrons to flow through the crystal and out of the other contact. What the physicists saw on the oscilloscope when they applied a voltage to the gold contact was a jump in the amount of current passing through the crystal and out of the other contact. They had made a solid-state electronic switch that exploited the ability of electrons in solids to leave their atomic anchors and instead become part of a river of electrons with no allegiance to specific atoms. A colleague would later suggest that they call the device a transistor.

The electronic amplification that went on inside a transistor meant that solid-state devices made out of crystals could serve the same function as the millions of comparatively large and energy-hungry vacuum tubes that had been the heart of electronic technology for decades. Achieving the amplification effect no longer required the heating of a metal filament inside of an evacuated glass tube to get electrons moving: transistors did the job using much smaller currents and in a far tinier space.

Transistors turned into a godsend for the American Telegraph and Telephone Company, the parent company of Bell Telephone Laboratories. AT&T managers had been growing mightily anxious about how they would be able to meet the growing communications demand of the nation if they had to rely on vacuum tubes for the millions of switches the growing network needed.

After rescuing mass telephony, transistors strode forth to lead the continuing microelectronic revolution, spurred on by the contributions of a diverse army of researchers. It may have been physicists who discerned the quantum mechanical basis for transistor action, but the microelectronics revolution that began to emerge in the 1950s and 1960s hinged on the work of metallurgists, chemists, and others who developed techniques for making crystals of unprecedented purity and perfection and for miniaturizing solid-state circuitry.

The multidisciplinary effort that developed transistors into a practical technology provided a preview of the future of technological innovation. Peering ever deeper into the substrate of matter, science discovered that materials as diverse as steel and plastic arise from a common ground that requires cooperative cultivation by a variety of scientific researchers. Their combined efforts continue to yield a rich harvest of knowledge and new technologies.

PART II

the *Birth* of a Superdiscipline

As the twentieth century reached middle age, the era of trial and error in materials-related research was reaching a point where the law of diminishing returns became operative. The days were over when an individual working solo stood much of a chance of coming up with dramatically better steel alloys, or magnetic material for motors, or a semiconductor for switches. To people running powerful countries, this was an alarming reality. One big lesson from World War II was that staying on the strategic cutting edge of warfare technology would mean having a reliable supply of research teams that could create materials with unprecedented combinations of properties.

The existing pantry of materials had sprung from a multitude of technical traditions going on in separate but parallel lines, some extending back thousands of years. If so many new materials were sprouting from such a disjointed approach as they were by the first half of the twentieth century, what might come out of a concerted effort to alloy the specific strengths of each materials tradition into a more systematic superdiscipline?

A yet-to-be-recognized multidisciplinary materials research culture indeed had been in the midst of forming itself. The seed crystals of the culture were set down in places like Los Alamos, New Mexico, during the Manhattan Project, at Bell Telephone Laboratories in Murray Hill, New Jersey, at DuPont's polymer research group in Wilmington, Delaware, at General Electric's research labs in Schenectady, New York, and a handful of other industrial labs. There also were a few seeds of the new culture at academic venues like the Massachusetts Institute of Technology.

In these places teams of researchers with different training and skills were churning out new ceramic, polymeric, and metallic materials every year. These were places where physicists, chemists, mechanical engineers, electrical engineers, and others with specialized skills discovered how badly they needed one another's technical strengths. However, these pockets of interdisciplinary synergism did not carry sufficient momentum to trigger a more general trend.

If a new materials research culture was going to thrive, something more had to happen. Even as the overall U.S. scientific community was ballooning in the 1950s, there was still no single field of research known by something like "materials science and engineering." There were metallurgists. There were ceramists and glass specialists. There were polymer scientists. There were solid state physicists adept in semiconductor science or magnetic materials. Any sense of unity among these various categories of researchers was idiosyncratic, not systematic.

It would have to become systematic, however. The American future in military arenas, the space age, jet aircraft (which were only experimental during World War II), and semiconductor science could not be assured by masters of any one discipline. It would take collaborations. Masters of different disciplines would have to transcend their traditional differences to forge more powerful partnerships. For the sake of national security—always one of the most powerful drivers of action and mobilizers of resources—a new field of materials research was more than a luxury: it was a necessity.

Space exploration called for spacecraft that could reenter the atmosphere intact and deliver warheads to Cold War adversaries, which in turn required the development of refractory materials that could take the punishment of searing rocket thrust and the frictional heat of atmospheric reentry. That would take chemists and ceramic engineers who knew how to make new rugged substances. It would take physicists who understood how heat and friction moved through and altered matter. It would take engineers who could see how the weight, stability, mechanical, thermal, and other properties of candidate materials might or might not work into the overall system of, say, an intercontinental ballistic missile or a manned space vehicle.

Jet engines, which produce far more thrust than propeller planes powered by internal combustion engines, demand new kinds of metallic alloys that could take the much higher heat and stresses. There are no shoulders in the jet stream. The added thrust came only when rapidly spinning turbines could compress air and fuel into the engine's combustion chamber. Alloys that could remain strong while spinning furiously at temperatures that brought all known metals perilously close to a taffylike state would be the only way to feel confident that the turbine blades of a jet engine would not fail catastrophically. It would take metallurgists, aeronautical engineers, physicists, and other specialists to invent the necessary alloys, superalloys, and manufacturing processes.

Solid state semiconductor research and development required the finest minds in theoretical physics to reveal how electrons move in a crystal's regimented labyrinth of atoms, chemists who could develop ways of producing the world's purest crystals and electrical engineers who could work these crystalline semiconductor devices into electronic circuits useful for radios, weapons, telephone switching, and other technologies.

Postwar optimism about harnessing atomic energy for the peaceful purpose of generating electricity led to the fanciful notion that electricity could become too cheap to meter. But making any electricity from the energy of nuclear fission would entail managing the eroding, embrittling, and otherwise adverse consequences that high amounts and long durations of radiation have even on inanimate materials. In 1956 a high-level Atomic Energy Commission advisory board even made a prescient recommendation for addressing the many materials issues that go along with harnessing nuclear energy for power generation. The board recommended establishing "Materials Research Institutes" in which no single material would be the focus of study, nor would one single traditional discipline dominate.

Another top-level advisory and planning organ that emerged at this time was the White House's Coordinating Committee on Materials Research and Development (CCMRD). Its very name reflects a conceptual unification of materials. Its members were the heads of materials-related research in each of the federal agencies involved in science and technology. The charge of the CCMRD was to discuss strategies that would keep a steady flow of essential innovations in materials available for national security. Their recommendation was to create academic environments in which materials problems could be approached using tools and theories from whatever scientific or engineering discipline was pertinent.

The message that improving existing materials and creating new ones had become a skeleton to modern technology development was spreading throughout the federal science and technology community after World War II. The Office of Naval

Research was hearing from its own Solid State Sciences Advisory Panel that materials research could address many pressing issues, including the sometimes catastrophic cracking of the welded-steel hulls of hundreds of World War II Liberty ships, the corrosion of metals in the salty ocean environment, and propulsion at sea. The Air Force conducted a study in 1957 that recommended the creation of a National Materials Laboratory much like the Materials Research Institutes that the Atomic Energy Commission had envisioned. In such a place, the development of advanced aerospace materials for jets could be conducted in a novel interdisciplinary environment better equipped to solve the new materials challenges.

In 1958 these and many other lines quickened toward an historical convergence. On March 18 President Eisenhower's science advisory committee handed him a short background paper that articulated the sea-change in the field of materials. "Unique environmental conditions associated with rocket propulsion, nuclear reactors, space flight, and vehicle re-entry have established the need for materials which are not currently available," the paper stated.[1]

The key to the new materials needed was to be found in the kinds of scientific advances that led to the transistor and to an understanding of precipitation hardening in metals. "There is a general feeling . . . that such advances in sciences as these can lead to a technology of materials engineering quite different from the metallurgy, ceramics, and polymer technology of today," the paper stated. "But to achieve this result it will be necessary to begin to relate the new fundamental knowledge of matter to the behavior of highly complex materials. This in turn will require scientists and engineers from many different disciplines—organic chemistry, physical chemistry, metallurgy, and solid-state physics—to associate their special knowledge and different points of view." Therein lies a call at the highest governmental level to form a new field of materials science and engineering.

The very same call was being heard on the local level as well. Just weeks before the president's science advisory committee members sent their short manifesto to the President, Morris Fine, then a young professor in the Department of Metallurgy at Northwestern University in Evanston, Illinois, had written a memo to the school's academic dean. He and his two departmental colleagues wanted to change the name of their department to the Department of Materials Science. There was no precedent in the academic world for the metallurgy professors' request.

Fine made his case for the departmental name-change in a letter to the dean. "In many fields the application of scientific discoveries to technical practice has been limited almost exclusively by the problem of the materials to be used," Fine wrote. "This is strikingly demonstrated by the currently important fields of missiles and nuclear energy, where the scientific principles have been known for a decade or longer. In these fields solution of the materials problems is the key to rapid development." Solving known problems is only the half of it, Fine continued. "When materials problems have been solved, new properties of a particular material discovered, or when new materials have been found, whole fields of technological importance often are opened up."[2]

As for what the departmental name change would actually mean in practice, Fine had this to say to the dean.

> Traditionally the field of material science has developed along somewhat separate channels—solid state physics, metallurgy, polymer chemistry, inorganic chemistry, mineralogy, glass, and ceramics technology. Advance in materials science and technology is hampered by this artificial division of the whole science into separate parts. . . . The advantages of bringing together a group of specialists in the various types of materials and allowing and encouraging their cooperation and free interchange of their ideas would indeed be very great.

The problem was that metallurgists, physicists, chemists, engineers, and other materials researchers residing in different disciplines and departments all had traditionally kept to their own. Each had developed different ways of thinking, published their work in different journals, read different texts, used different jargon, and essentially lived under different research cultures with different mores and measures of accomplishment. At Northwestern, the Materials Science Department was supposed to be a different animal, a seamless hybrid of many disciplines with a long history of parallel development and almost no history of joint development.

Fine, who never left Northwestern after joining in the late 1940s, had the right kind of experience to back his then audacious plans to open up the country's first academic materials science department. He was one of the few scientists at the time who knew firsthand just how powerful and fruitful multidisciplinary research can be. The most powerful weapon and technology humanity ever had produced came out of the Manhattan Project at Los Alamos where Fine worked on the plutonium metallurgy team under Cyril Smith. After the war, he was recruited directly from Los Alamos to Bell Telephone Laboratories for the explicit purpose, he recalls, of bridging the physics and metallurgy researchers there. One of the recruiters was William Shockley, who, with John Bardeen and William Brattain, won the Nobel Prize in 1956 for inventing the transistor and who later seeded what became known as Silicon Valley.

Fine's own first efforts at Bell Laboratories focused on the crystallization of germanium, which was the flesh of the first transistor, and on silicon, which became the semiconductor material that changed the world as much as any material in history.

Fine's argument to his dean was persuasive. On December 29, 1958, the Department of Metallurgy officially changed its name to the Department of Materials Science.[3] A new research tribe was born. Northwestern University became one of the first academic departments to deliberately train students not only as metallurgists or ceramic engineers or polymer chemists but as materials scientists.

Meanwhile, the six-hundred-word paper by President Eisenhower's science advisory committee had set a policy machine in motion. Just as Northwestern's materials science department began its life in late 1958, all of the federal advisory committees and bureaucratic report writing began bearing fruit. At the focus was Herbert York, then the director of the newly formed Advanced Research Projects Agency (ARPA) with a Cold War charge to make sure that the country did whatever it would take to develop intercontinental ballistic missiles. In a high-powered meeting between officials from the Atomic Energy Commission and the Coordinating Committee for Materials Research and Development, York was convinced that the creation of well-supported interdisciplinary laboratories (IDLs) for materials research at universities was a sound course for the government to take and for ARPA to fund.

For one thing, the creation of such an R&D infrastructure could only help ARPA's mission to build ICBMs as well as future systems deemed important for national security. After all, planners envisioned ICBMs as heaving nuclear warheads through the atmosphere into outer space, shuttling the weapons halfway around the world to some target, and then plowing again through the atmosphere intact to deliver their destructive cargo. If anything, this would be an intensively materials-dependent challenge.

Within a year of taking on the responsibility, the first three IDL contracts had been drawn up and signed by Northwestern University, Pennsylvania State University, and Cornell University. All three already had adopted the interdisciplinary mind set to some degree, so the IDL seeds were sown on fertile soil.

The mission of these labs was about as comprehensive as

lab missions could get. The original IDL contract stated the objective as "furthering the understanding of the factors which influence the properties of materials and the relationships which exist between composition and structure and the behavior of materials." The deliberate generality of the contract's language helped bring all classes of materials—metals, ceramics, polymers, and so on—within a single conceptual framework.

In the early 1960s nine more ARPA contracts, three Atomic Energy Commission (AEC) contracts, and two NASA contracts expanded the nation's interdisciplinary materials research infrastructure throughout the country. By 1972, when the National Science Foundation took over the IDLs from ARPA and renamed them as "Materials Research Laboratories," materials research had become a genuine field at many major universities, not just those with special government funding. Single-discipline departments that for decades had gone by names like metallurgy or minerals and mining were falling under broader-based materials departments.

As these laboratories were getting off of the ground, yet more institutional cornerstones for the new field of materials science and engineering were poured and cured. One of these was put into place in the spring of 1973 when the Materials Research Society (MRS) held its inaugural meeting at Pennsylvania State University. Rustum Roy, a ceramic materials expert at Penn State, founding member of MRS, and perennially one of the most influential and colorful members of the materials research community, recalled that "just under three hundred people, including Penn State faculty and students, attended."[4]

A humble beginning, to be sure, but the creation of professional societies is a hallmark for budding scientific disciplines because it provides a core from which its members can further develop a community sensibility. Since 1960, when those first IDLs were commissioned, the new breed of interdisciplinary

1938
First epoxy resin patent granted.

Teflon™, the product of the polymerization of tetrafluoroethylene, developed.

High-pressure polymerization of ethylene introduced.

W. H. Carothers' invention of nylon announced.

Polyvinyl butyral resin sheeting introduced, making automobile safety glass possible.

Thermosetting binders ("warm"-pressing ceramics) patented.

1939
Low-density polyethylene (LDPE) commercially produced.

Cellulose acetate molding compound and ethyl cellulose produced.

1940
Formica™ and poly(vinyl chloride) (PVC) produced.

Bomber noses of Plexiglas acrylic resin are thermoformed for war planes.

Film and sheet vinylidene chloride introduced.

Jet-molding machines developed.

Prototype plastic car developed, made of soybeans and wood fiber processed into plastic, offers improved impact resistance.

1941
Fiber-forming polyesters developed.

Polyethylene terephthalate (PET) invented.

Superconductivity demonstrated at 15 K.

1942
Polyacrylonitrile (PAN) fibers (Orlon™) commercially introduced.

First polyethylene bottle blown.

Thermoformed plastic blister packaging developed.

First working electromechanical digital computer developed.

1943
Saran monofilament extruded.

Free-blown acrylic aircraft canopy produced.

Silicone resins produced.

E-glass (electrical glass), a type of fiber glass, patented.

Injection blowmolding process patented.

1950
Commercial production of Teflon™ begins.

The International Standardization Organization establishes Plastics Committee.

Dacron™ polyester fiber for wrinkle-resistant, permanent-press clothing introduced.

Commercial production of titanium begins.

1951
Injection-molded model aircraft hobby kits produced.

Commercial production of high-purity, single crystal silicon begins.

Field ion microscope developed.

1952
Catalysts for low-pressure polymerization of ethylene developed.

High-impact, rigid poly (vinyl chloride) developed, making PVC pipe possible.

Fiberglass-reinforced polystyrene developed for military and commercial applications.

Mylar™ polyester film introduced, for recording tape, shrink-wrap packaging, and electrical insulation.

Basic oxygen furnace for steelmaking developed.

Silicon solar cell introduced.

1953
Polycarbonate (PC) invented.

First Dacron™ (PET) plant begins production.

Chevrolet Corvette uses 41 fiberglass reinforced parts, including 17 major assembly components.

Early large-scale integration circuitry developed by layering wafers of miniaturized electronic circuits.

Tungsten-nickel-copper composite first developed and used for radioactive substance containers.

1954
Propylene (PP) is polymerized.

Molybdenum-manganese metal coating patented for use in vacuum-tight, metal-to-ceramic seals.

FIGURE 10

Twentieth Century Stuff. This fifty-year time line beginning in 1938 shows milestones in the invention, introduction, applications, or commercializiation of various materials and materials-making processes.

THE DEPARTMENT OF THE INTERIOR, BUREAU OF MINES

1944

BT-15 airplane with fuselage fabricated in a fibrous-glass-reinforced plastic sandwich structure fuselage flown.

First low-pressure-thermosetting prepregs developed.

1945

Dielectric preheating introduced.

Zytel™, an extremely tough nylon molding and extrusion resin, introduced for use in gears and cams, sports equipment, valves and other products.

1946

Acrylic introduced in line of dentures and automobile taillight lenses.

Polyethylene appears in kitchenware.

Nylon zipper introduced.

Polyester developed.

ENIAC, the first all-electronic analog computer, introduced. Weighing about 30 tons, it required 150 kW to operate its 18,000 vacuum tubes and other devices.

1947

Vacuum melting technology developed.

Etched electronic circuit created.

Germanium transistor developed, beginning a trend toward miniaturization. Its first practical application was in hearing aids.

Epoxy introduced commercially in the U.S. as an adhesive.

1948

Acrylonitrile-butadiene-styrene (ABS), used to manufacture luggage, introduced.

1949

Screw plastication with automatic transfer molding developed, using a nonreciprocating screw.

"Doctor-blading" process of forming thin ceramics patented.

1955

High-density polyethylene (HDPE) developed.

Hypalon™ synthetic rubber introduced, offering advantages over rubber. Used in roofing, auto, home, and industrial applications.

Cronar™ polyester film base for x-ray and graphic art films developed.

Hot isostatic pressing technology developed.

1956

Polyphenylene oxide (PPO) discovered.

Lucite™ acrylic lacquers introduced.

1957

Widespread automotive use of urethane foam for seating and safety dashboards inaugurated.

Polyacetal resins developed, offering high strength with toughness of metal for use in hardware, auto parts, cams, rollers, and zippers.

Soviets launched Sputnik. The space race begins.

Silicon transistor developed, permitting devices to operate at higher temperatures and frequencies with greater stability.

1958

Lycra™ spandex fiber introduced. Stretch up to five times its length and returns to original length.

Ruby-crystal laser developed.

First commercial production of microalloyed columbium steels.

1959

Commercial production of acetal homopolymer begins.

Integrated circuit (IC) developed with multiple silicon transistor circuits on a single silicon wafer.

Molybdenum-wire-reinforced-titanium matrix composite developed.

1960

Ethylene vinyl acetate copolymers introduced.

Acetal copolymer used in automobile brake-cable pulley.

High-impact styrene furniture legs with through-bolt construction developed, starting trend to plastic furniture.

Glass-ceramic patented.

Amorphous metal alloy produced.

Synthetic diamond production begins.

Aircraft with fiberglass-polyester skin and paper honeycomb core first flown.

(continues)

Advantages of parallel metal grain boundaries for turbine blades demonstrated.

Effects of rapid solidification process first reported.

1961
Tape process of forming thin ceramics patented.

1962
Polyvinylidene fluoride introduced.

School furniture introduced, combining solid and laminated plastic in a tubular steel construction.

Transparent polycrystalline alumina patented.

Scanning electron microscope (SEM) developed.

Nickel-based superalloys using oxide-dispersion strengthening announced.

Gallium arsenide laser diode developed.

1963
Polyimides introduced, increasing the thermal endurance of thermoplastics to 750° F.

Union Carbide process invented for low-pressure structural-foam processing.

Nomex™ flame-resistant aramid fiber and paper introduced for protective clothing, high-performance hoses, high-temperature electrical uses.

Float process for making glass patented.

Sintered alumina abrasive grain patented.

1964
Polyphenylene oxide (PPO) components for appliances and electrical connectors produced.

Surlyn™ ionomer resins introduced, offering clarity of glass and toughness for use in food packaging, sporting gods, and automotives.

"Certi-Fired" thick film materials for electronic circuit miniaturization introduced.

1965
Polysulfone developed and introduced, finds commercial uses in electrical components.

Glass-reinforced styrene-acrylonitrile (SAN) copolymers appear in automobiles.

Clysar™ shrink film used in cold packaging applications.

Process patented for making dense impregnated, silicon-carbide articles.

Cobalt-rare earth magnets developed.

1972
Polyethersulfone used in aerospace and automotive applications.

Robotics used in the plastics industry as the first high-speed, machine-mounted, automatic part remover for injection is patented.

Imron™ polyurethane enamel introduced for automotive applications.

Sialons developed.

1973
Kevlar™ aramid fiber introduced, which, on a weight basis, is five times stronger than steel. Used in sporting goods, industrial products, bullet-resistant vests, automotives, aerospace, etc.

NASA Langley Research Center initiates flight-service program placing carbon/epoxy assemblies into service.

Superconductivity demonstrated at 23 K.

1974
Carbon-reinforced epoxy upper aft rudders introduced in MacDonnell Douglas DC-10.

Acrylic sheet, stiffened with reinforced plastic used for all exterior body panels in an automobile.

1975
Reaction-injection molded urethane held in place by a glass-reinforced polypropylene sheet retainer is used as the front end of the Chevrolet Monza.

1976
All-plastic jet designed.

Plastic microwave cookware becomes available to the consumer market.

Amorphous silicon solar cell introduced.

1977
Commercial production of linear low-density polyethylene (LLDPE) begins.

Polyphenylsulfone (Radel™) introduced.

First polymer demonstrating electrical conductivity is synthesized.

First space shuttle launched.

Fusion welding of silicon-carbide-reinforced aluminum demonstrated.

Neodymium-iron-boron magnet developed.

1983
Conductive composites for computer housings are developed as FCC regulations require shielding of plastic housed electronics components.

Alumina-silica-fiber-reinforced aluminum piston for diesel engines developed.

1984
Melt-processible liquid crystal polymers (LCP) introduced (Xydar™).

Five ship sets of composite horizontal stabilizers installed on the Boeing 737.

Plastic fuel tank for U.S. passenger car is blowmolded and treated with sulfonation process to control hydrocarbon permeability.

The twin-turboprop Avtek 400™ promotes the art of advanced composites.

Selar polyamide barrier resins introduced to blow-molded bottles used for containing chemicals and hydrocarbons.

Quasi-periodic crystals discovered.

1985
Bexloy™ engineering resins introduced for low-weight, high-strength auto body parts to replace metal.

Alumina-fiber-reinforced aluminum experimental squeeze casting introduced.

Graphite-fiber-reinforced magnesium vac-assisted investment casting introduced.

Electronic discharge machining of metal matrix composites demonstrated.

1986
The Voyager, a composite aircraft, flies around the world without refueling.

Superconductivity demonstrated at 39 K.

Large billets of both silicon-carbide-particle-reinforced aluminum and silicon-carbon-wire-reinforced aluminum produced.

Alumina-fiber-preform-reinforced aluminum introduced.

1966
Noryl™, a modified polyphenylene oxide (PPO), introduced.

Uniloy machine for high-density polyethylene blowmolded milk bottle production introduced.

Acrylonitrile-butadiene-styrene (ABS) used on exterior surfaces of helicopter.

Kapton™ polyimide film, offering resistance to moisture and extreme temperatures, is developed and becomes valuable for aerospace use.

Fiber optics developed.

1967
Polyarylsulfone introduced as Astrel 360™.

Permasep™ permeators introduced as an economical way to desalt water.

The first composite aircraft to receive FAA certification, is built.

Single crystal turbine blade is demonstrated.

1968
Polyphenylene sulfide (PPS) introduced as Ryton™.

Chrome plated polypropylene used in automotives.

Manned moon landing introduces plastic components to new applications.

Riston™ photopolymer dry film resists are introduced as light-sensitive polymer films for printed circuit board production.

Very large-scale integration (VLSI) electronic circuitry developed.

1969
Corian™ introduced as a stain-, scratch-, and burn-resistant non-porous material for countertops, sinks, etc.

Superconductivity demonstrated at 21 K.

1970
Boron-epoxy horizontal stabilizer on the F-14 represents the first advanced composite part produced that was designed as a composite part and not as a substitute for metal.

1971
Mechanical alloying developed.

Metal injection molding developed.

Zinc oxide varistor patented.

1978
Polyetherether-ketone (PEEK), a high-temperature resistant material, becomes available for aerospace and computer applications.

Polyarylates introduced.

The practical use of optical fibers begins.

1979
Volume production of plastics surpasses that of steel.

The Gossamer Albatross, made in large part of Mylar™ polyester film, becomes the first human-powered aircraft to cross the English Channel.

Rynite™, a heat-resisting, very rigid polyethylene terephthalate (PET), introduced for use in electronic components, furniture, lighting fixtures, industrial machines, and automotives.

Tungsten-fiber-reinforced ferro-chromium-aluminum-yttrium used experimentally for jet engine rotor blades.

1980
Acid-leaching process introduced for producing 99.6% to 99.9% silica fibers that resist devitrification up to 1,370° C. Used as insulation for the Space Shuttle.

1981
Pyralin™ polyimide coatings used as insulation in semiconductor chips.

Superconductivity used in magnetic resonance imaging medical equipment.

1982
Modified polyimide (polyether-imide) (Ultem™)introduced for use in high-performance fiber optic components, coextruded food packaging, and advanced composites.

High-purity polycarbonate resin for manufacture of compact disks ("CD's") introduced.

Jarvik-7 artificial heart designed, made largely of plastics, supports patient for 112 days.

Molten metal oxidation method used to produce net shapes.

Commercial production begins of alumina-silica-fiber-reinforced aluminum pistons by squeeze casting.

Boron-fiber-reinforced aluminum used in sports equipment.

Water jet and laser cutting of metal matrix composites demonstrated.

1987
Superconductivity demonstrated at 98 K.

Silicon-carbide-particle-reinforced aluminum, optical-grade material developed.

Discharge welding of silicon-carbide-fiber and wire-reinforced aluminum demonstrated.

Silicon-carbide-particle-reinforced aluminum formed by standard forming technique.

Silicon-carbide-fiber-reinforced titanium and silicon-carbide-fiber-reinforced titanium-aluminum, experimental metal matrix composites developed for temperatures near 2000°F.

1988
Superconductivity demonstrated at 125 K.

Polymer semiconductors developed.

1989
The first all composite business jet, Beech Starship, goes into commercial production.

Development of high-purity, low-defect indium phosphide single crystals, which has high potential for use in space-based solar cells.

materials researchers have fanned out to nearly every university, hundreds of government laboratories, as well as thousands of small and large corporate research laboratories and plants. The tribe has spread throughout the land. This has been true not just in the United States but throughout the world.

To get an idea of the growth in the field since the MRS was founded, consider that in 1986, when the Materials Research Society began publishing the *Journal of Materials Research*, MRS membership had expanded from its original 1972 membership of 215 to over 4,000. By 1996 MRS membership was over 12,000. And MRS is only one of many materials-oriented membership organizations. A more comprehensive measure of the scope of the field of materials science and engineering comes from the membership of the Federation of Materials Societies (FMS), an umbrella organization representing more than a dozen large and small materials-related societies and association. FMS claims about 750,000 members. The American Chemical Society (ACS) and its 150,000 members is among the larger organizations on the roster. The smaller FMS members, whose constituencies number in the thousands, are more specialized and go by names like the American Welding Society and the National Association of Corrosion Engineers. There are dozens of other materials-related societies outside of the FMS, so that 750,000 is almost certainly a conservative estimate.

These materials researchers already have succeeded in changing the way humanity discovers new materials, incorporating trial and error but advancing beyond it to a new approach to materials based on the way their structure and anatomy relate to their properties and technological performance. Instead of always limiting designs to the known performance limits of available materials, engineers have begun to take the audacious step of specifying the new materials they need for their ever more ambitious technological designs. If a desired material doesn't exist, then materials researchers ought to be able to figure out how to make it.

Among the historic precedents for this approach is the Manhattan Project, whose charge to create nuclear weapons was tantamount to a challenge to develop new materials like the radioactive metal plutonium. Another came a few years earlier in the form of the government's "GR-S Program" whose goal was to create almost overnight a massive industry for making hundreds of thousands of tons of synthetic rubber to replace embargoed supplies of natural rubber. President Kennedy's successful Apollo program confirmed this philosophy again in the 1960s. These projects showed how a combination of basic materials science and engineering research had begun a synergistic spiral toward ever greater control over the material world.

Despite these gains, materials science and engineering remains an obscure outpost of the country of science. People on the street have heard about biologists, chemists, and physicists—but materials scientists? In 1990 this unseen research community made its boldest bid to earn some name recognition by producing a book-length report, *Materials Science and Engineering for the 1990s.*[5] Known as the "Blue Book," it makes the case that the U.S. economy is so heavily dependent on advances in materials that research in the field ought to become a national priority. "Without new materials and their efficient production, our world of modern devices, machines, computers, automobiles, aircraft, communications equipment and structural products could not exist," the MS&E report declared.

This is really a matter of old news striving for contemporary attention. The ancient Phoenicians knew it, traveling to southern England for the tin they needed to manufacture bronze weapons. It was clear to John Roebling in the nineteenth century, who saw in steel the possibility of suspending a beautiful bridge across the East River from Manhattan to Brooklyn. It was clear to Wallace Carothers and his managers at DuPont

FIGURE 11

Metallic Forests. These dendritic (treelike) microstructures formed during a welding process involving a nickel-based alloy, the kind of mechanically strong alloy used to make the turbine blades in the engines of airliners. Symmetrical structures of this kind are produced as the already solidified portion of a weld forms "side branches" while penetrating into the still molten portions of the weld. Revealing structural details of this sort is a critical part of the quest to improve the performance of materials, including welded metal parts. In this scanning electron microscope photograph, the hot liquid phase has been expelled to reveal the dendritic structures.

SEM MICROGRAPH COURTESY OF S A DAVID , L A BOATNER, AND R STEELE OF OAK RIDGE NATIONAL LABORATORY, OAK RIDGE, TENNESSEE

in the 1930s, who saw a technological and commercial bonanza in polymer molecules. It was a given for aerospace engineers at Northrup and other aerospace companies in the 1980s and 1990s, who developed more than nine hundred new materials for the impossibly expensive B-2 *Stealth* bomber.

But today's avant-garde is often tomorrow's establishment; patiently pushing the capabilities of technology's flesh and bone, the obscure visionaries of contemporary materials science might well be the architects of the future's commonplace miracles.

FOUR

MATERIALS *Research* COMES OF AGE

In 1960, when the field of materials science and engineering was born, a silicon chip the size of a fingernail harbored one solitary transistor. Much smaller than the vacuum tubes that were the basic unit of computers at the time, a chip amounted to one switch, one on-off gate. But to make a computer, you still had to link a very large number of fingernail-sized chips together.

Ten years later, the same amount of silicon real estate held nearly ten thousand transistors. The component density was up to several hundred thousand by 1985. By the mid-1990s, microelectronics manufacturers were packing millions of transistors and other electronic components onto a single chip. A single chip today, running on a few lightbulbs' worth of electricity, is far more powerful than the room-size, electricity-guzzling computers of the 1950s that required full-time staff just to keep replacing the machine's vacuum tubes, which were constantly burning out.

Each chip has become a micrometropolis of circuitry whose

individual components could sit within a biological cell. When you make a graph that shows how the number of transistors on a fingernail-size piece of silicon crystal has increased over the years, you get an upward sloping curve that gets steeper and steeper. If you were to walk the curve, it would be like walking up a ski ramp.[1]

This feat of microelectronics fabrication would not have happened without materials scientists. Crystal growth experts were needed to make silicon crystals nearly devoid of the tiny defects and contaminants that had made the first transistors too unreliable for military planners to rely upon. It took polymer scientists to develop the photolithographic materials required for patterning circuitry onto the silicon wafers cut from those crystals. It took the expertise of ceramic engineers to develop materials to entomb the delicate microcircuitry so that it can be handled and put into products.

Since the birth of materials science and engineering, materials researchers have consistently improved materials that have in turn enhanced the performance of the technologies that use them. If you plot the progress of some measure of performance (engineers call such things figures of merit) over time for each technological function like electrical conductivity or optical transparency, you get another ski-jump ramp. The steepness of the ramp really takes off in each case when a material receives the sobriquet of "advanced material," which denotes an

FIGURE 12

The Ski Ramps of Progress: The graph shows how the progression from wood and stone to various metals to the advanced polymers (aramid fibers) and ceramics (carbon fibers) of today has yielded materials that grow ever stronger while using ever smaller amounts of mass. Similarly accelerated rates of improvement in many material properties in recent years yields similar ski-ramp curves of progress.

REPRINTED WITH PERMISSION FROM *MATERIALS SCIENCE AND ENGINEERING FOR THE 1990S MAINTAINING COMPETITIVENESS IN THE AGE OF MATERIAL* (WASHINGTON, DC THE NATIONAL ACADEMY OF SCIENCES PRESS)

unprecedented leap in superiority over previous products. (See Figure 12.)

Take the so-called "strength-to-density" ratio of materials, a measure of how much of a material it takes to support a given load. The higher the ratio, the less material you need. The strongest thing going in Antoine Lavoisier's time in the late eighteenth century was good cast iron. A cast-iron rod with a square cross section one inch on a side and weighing 4 pounds per foot could support up to 50,000 pounds (25 tons), or about 330 adults. Cast iron's strength-to-density ratio actually is not much better than bronze or, for that matter, wood or stone. In the nineteenth century steel started to push the strength-to-density into ski-jump levels. But not until the 1950s—when scientists invented polymer-based composites reinforced with

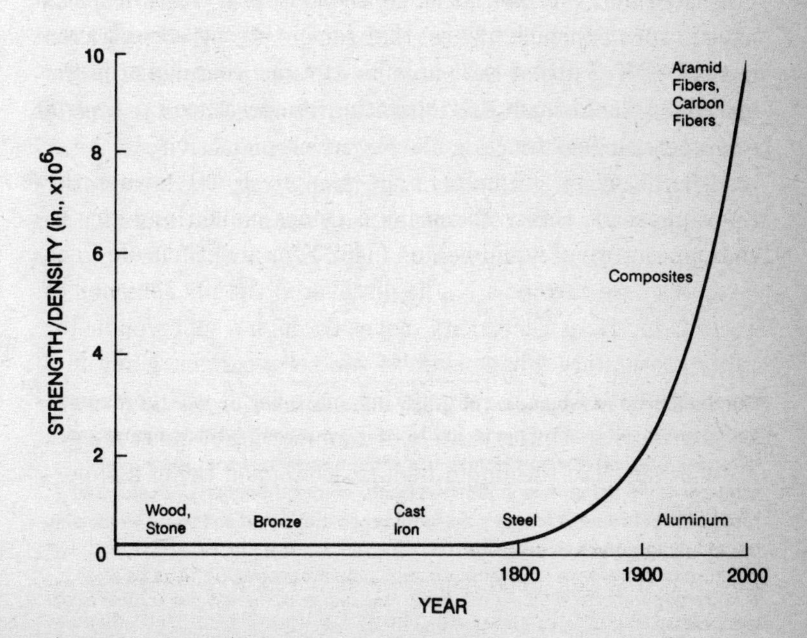

glass and other ceramic fibers and high-strength polymers such as Kevlar (a polyaramid fiber)—did the ramp get really steep. Modern high-strength materials—say, a cable made of the carbon fibers in military aircraft or high-end golf clubs—could support the same three hundred people as Antoine Lavoisier's cast-iron rod, but the modern cable would only have to be one third of an inch on a side and one sixtieth as heavy. It also would cost a whole lot more.

The historical trend of another figure of merit—the highest temperature at which an engine can operate without its metal parts weakening, melting, or otherwise giving up the ghost—also reveals the technological meaning of materials research. The higher the operating temperature, the greater the engine's efficiency. In 1900 steel steam engines operated at the boiling point of water, about 212°F. In the 1930s air-cooled airplane engines were running about 380°F. Since then metallurgical researchers at aerospace engineering companies have developed superalloys (grown as single monolithic crystals of metal in the shape of turbine blades) that remain strong enough even above 2,000°F. Putting heat-immune ceramic coatings atop the superalloys could push that operating temperature up several hundred more degrees.

The litany of ski-jump ramps goes on. In 1911 when the Dutch physicist Heike Kamerlingh Onnes chilled mercury to the temperature of liquid helium (–452°F, or about four degrees above absolute zero or 4 K), he discovered the phenomenon of superconductivity. Electricity moves through a superconductor with no resistance whatsoever. If you set a current going in a ring made out of a superconducting material, it will go around the ring forever. Unfortunately, if Kammerlingh's mercury got any warmer than the liquid helium, it stopped being a superconductor.

For the next seventy-five years researchers searched for materials that remained superconductive at warmer temperatures, since refrigerating anything to liquid helium temperatures is costly and difficult. They did, but the warmest superconductor they could find—a niobium-germanium intermetallic material discovered in the 1970s—still lost its superconductivity at –417°F (23.3 K).[2]

The shallow upward slant of the previous seventy-five years became a ski jump in 1986, when two scientists at IBM's Zurich Research Laboratory started a research wildfire with their discovery of a ceramic material—a barium lanthanum copper oxide—that remained superconductive at the unexpectedly warm temperature of about –396°F (35 K). They and others soon learned that the barium lanthanum copper oxide was just the first of an enormous class of superconductive ceramic materials. By 1996 a growing army of materials researchers discovered ceramic superconductors with transition temperatures as high as –216°F (135 K). The search for yet warmer superconductors continues.

There is a catch. The ceramic superconductors are more like rocks than like the earlier metallic superconducting materials. It is hard to make a rock into a wire, which is the kind of shape you want to work with for electrical applications. Nonetheless, researchers have made experimental cables and electronic circuits made with the high-temperature ceramic superconductors.

Perhaps the most dramatically steep and rapid improvement in a material is the transparency of glass. As late as 1965, no light sent into a thousand-meter length of glass fiber drawn from the best glass available would make it to the end. At the time glass fiber was far more useful for making insulation or as part of a polymer-glass composite for things like boat hulls than it was for piping light. Since then, however, materials researchers have developed ways of manufacturing endless miles of optical fibers so devoid of the material imperfections that used to darken the light that almost all light put in one end will shine through the other.

The transparency of optical fibers has increased by one

hundred orders of magnitude—that is, by a factor written with a 1 followed by one hundred zeros. The hair-fine fibers strung across late-model transoceanic optic fiber cables can carry one-hundred and twenty thousand phone calls at once. Today's best optical fibers can relay data at a rate of billions of bits per second, the equivalent of an entire Encyclopedia Britannica every second. This is the kind of ski-jump ramp that leads to talk about information highways.

There are many more ski-jump ramps that show how far materials researchers have come. Today, there are neodymium-iron-boron-based magnetic materials a hundred times stronger than the magnetic steel alloys used at the turn of the century. Every increase in the magnetic field leads to smaller and more efficient motors for everything from heating systems to windshield wiper blades. There now are tool materials such as cubic boron nitride and tungsten carbide coated with synthetic diamond that enable manufacturers to cut, shave, and mill materials into products a hundred times faster than the tool steels of the early century. Tools that work a hundred times faster can produce a hundred times more product in the same time or the same amount of product in one one-hundredth the time.

The "superdiscipline" of materials science and engineering (MS&E) has its ski prints all over these ski-jump ramps. Evidence of the consequences is manifest throughout today's high-tech society. Whenever materials researchers have introduced harder, stronger, lighter, or more electrically conductive materials—or for that matter, any affordable material that is better in some way than all available materials—more capable technologies follow. With better materials come faster jets, higher-capacity communications, more powerful computers, more energy-efficient cars and appliances, longer-lasting and more capable medical prostheses, and more woeful and devastating weapons, to name a few consequences. (See Figures 13 and 14.)

The rocketlike acceleration in the performance of materials in the past four decades is due to an unprecedented three-way bootstrapping: the ability to reveal the structure of materials and how they form; the ability to form materials into specific microarchitectures; and the ability to understand, model, and simulate material phenomena using math and computers—in short, analysis, synthesis, and theory, respectively.

Just as diversity exploded in the biological realm during the Cambrian revolution, so the diversity of analytical tools and methods has proliferated in materials science in recent decades, giving researchers a seemingly unobstructed view of the anatomy, formation, and transformation of materials to atomic levels of detail as they witness processes at time intervals as short as quadrillionths of a second.

As each member of this triumvirate—analysis, synthesis, and theory—has diversified and ramified, it has boosted the sophistication of the other members. The microelectronics revolution has raised the water level of all three. Computers and microelectronics have enabled researchers to control their instruments with far more finesse, to gather far more data, and to make sense of data far more quickly; and they have promoted better control of the reactions and processes by which ingredients become materials. Manufacturers constantly test theorists' ideas in an iterative loop that benefits both theory and practice. The result: materials science and engineering has achieved a mastery of the material world that was unthinkable in 1960, when the field was born.

Here is a brief survey of some of the analytical methods, synthetic methods, and theoretical and computational tools that have fueled this ascendance.

Analysis. Henry Sorby could use optical microscopes to see structural detail down to the point where he could distinguish grains the size of cells or thereabouts. By 1912 physicists were using X-ray crystallography to examine details of the average crystalline structure of metals and other solids. In the 1930s the advent of electron microscopy opened vistas between optical microscopy and X-ray crystallography. In 1951 a scientist at the University of Pennsylvania saw some of the first human "glimpses" of individual atoms with his invention—the field ion microscope. It worked by using intense electric fields to yank atoms off of the end of a stylus-shaped sample and then accelerating the atoms into a detector that could reveal the relative locations of the liberated atoms.

One variation of the field ion microscope essentially shoots movies of a sample as atomic layers sequentially evaporate off and hit a detector. After enough layers have come off, a three-dimensional portrait in atomic detail emerges the way a CAT scanner can construct a three-dimensional image of the inside of a person from many two-dimensional X-ray slices.

Along the way, too, have come various spectroscopies. The first of these techniques goes back to the nineteenth century, when scientists were first learning to harvest chemical information from the colors of light emitted from or absorbed by their samples. Since then the battery of spectroscopic instruments and methods has grown to astounding versatility. All of them

---

The Sculpture of Ceramic Structure. Modern synthesis and processing techniques can produce materials in an endless variety of forms. Top right: The scanning electron microscope photograph (SEM) depicts a particularly sculptural feature from a sample of zinc oxide crystals produced by a high-temperature growth technique known as "chemical vapor transport." The process is akin to the formation of ice crystals when the moisture in your warm breath blows over a wintry window pane. Zinc oxide ceramics are important electronic materials that are used for electrical surge protection in devices ranging from large-scale power distribution systems to consumer electronics such as televisions and home computers. Bottom right: This SEM photograph shows a porous microstructure in potassium tantalate ceramic, whose underlying crystal structure, known as a perovskite, is a building block for a number of important compounds, including ferroelectric ceramics that can store high-quality images and high-temperature superconductors.

FIGURE 13 L A BOATNER AND R STEELE, OAK RIDGE NATIONAL LABORATORY, OAK RIDGE, TN

FIGURE 14 L A BOATNER, J O RAMEY, AND R STEELE, OAK RIDGE NATIONAL LABORATORY, OAK RIDGE, TN

FIGURE 13

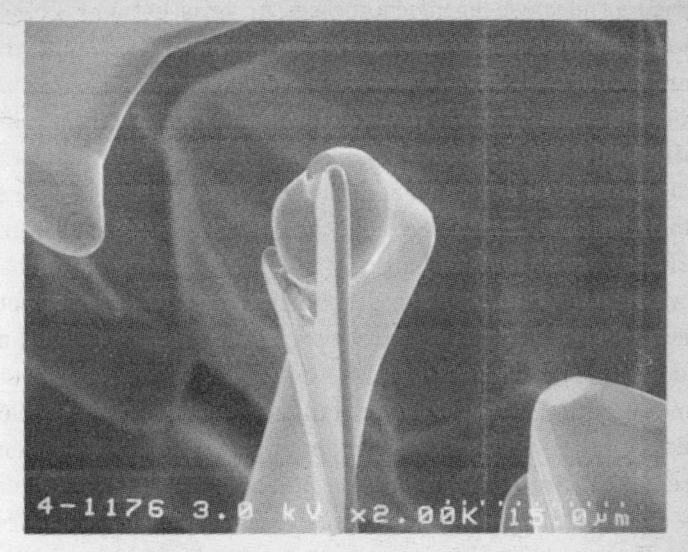

FIGURE 14

measure the wavelengths, intensities, or variations in visible or invisible light, or in energetic particles such as electrons and neutrons going in or out of samples.

There is femtosecond laser spectroscopy, in which laser pulses in the vicinity of a million billionth of a second can prod and probe the detailed events that occur as atoms or molecules approach one another, react, and then diverge. The result is like a slow-motion picture of a great play or blunder in a football game, only in this case the detail helps materials researchers know what structures are worth trying to emulate or avoid as they try to make new materials or improve old ones.

There is low-energy electron loss spectroscopy, in which changes in the speed and direction of slowly moving electrons directed at samples reveal subtle motions and dynamics occurring at the surfaces of materials such as industrial catalysts.

There is laser ablation mass spectrometry, which can take a speck of sample barely visible to the naked eye, train a laser onto a tiny subregion of that speck, and then blast part of that region into a molecular plume. The mass spectrometer then can weigh and thereby chemically identify those plume components by monitoring how little or how much they deflect in a magnetic field. These kinds of molecular identifications are part and parcel of developing new materials and the processes for making them.

Since 1981 a new family of even more powerful microscopes, the so-called scanning probe microscopes, have been routinely resolving blurry ghosts of individual atoms so that the tiniest of imperfections—a single atom out of place in one crystal grain or on a semiconductor chip—can be discerned. The scanning tunneling microscope (STM) was the mother of all scanning probe microscopes. Invented by Gerd Binnig and Heinrich Rohrer in Rüschlikon, Switzerland, at the IBM Zurich Research Laboratory where the first ceramic high-temperature superconductor was discovered, their STM, like an old-fashioned record player writ small, permits imaging of the atomic and molecular landscapes of surfaces with unprecedented ease and clarity.

The principle behind an STM is to bring an extremely fine stylus close to an electronically conductive surface so that the electron tunneling can occur. Even though there is no direct contact between the stylus and the surface, electrons tunnel between the two. The quantum current dramatically strengthens or weakens upon the slightest difference in the distance of the stylus's tip from the surface. As the computer-controlled stylus effectively sweeps back and forth over the surface like a blind man's walking stick on a sidewalk, the computer keeps track of the changing tunneling current. This electrical record then can be reconstructed into a three-dimensional image of atomic resolution complete with operator-chosen false coloration. The gorgeous computer-generated pictures from these exercises have changed the look of scientific journals.

Each of these instruments, as well as a growing roster of variations for many of them, provides analytical access to almost any structural level of any material before, during, and after the material's synthesis, processing, or service life. The analytical arsenal available to materials researchers today is nothing short of miraculous. Yet the power to see details of materials still accelerates. There is a small army of researchers with a special brand of machismo that expresses itself as a drive to develop instruments and techniques that push this analytical envelope further and further. For them life is like a technical "name that tune" game, only the boast is more like "I can name that substance with one molecule," or "I can determine the distribution of grain sizes and point defects in that mite-size sample."

Synthesis. Even if you know the precise location and activities of all the atoms in a material, it won't get you a new polymer or a new metallic alloy. That takes putting things together, not taking them apart: it means obtaining raw materials or synthetic ingredients and then processing them into real materials. Just as a cadre of scientists have brought analytical instrumentation and capability to virtually the ultimate level, so too have makers of chemicals and materials.

Metallurgists have learned how to use specific heat treatments to control the way the atomic components of molten metal aggregate, separate, and fall out into a nexus of crystalline metal grains, ceramic particles, voids, and other features that constitute a particular alloy with unique characteristics.

Chemists have gotten to the point where they can probably find a way to synthesize any molecule that is physically possible. Some of the most dramatic displays of molecule-building prowess have come from organic chemists seeking to duplicate in the lab the complex chemical structures that living cells have brought to an enviably high art. Materials scientists and engineers need to think beyond that. Not only do they need to have suitable atoms and molecules, but they also need to find ways of assembling these into ever larger structural arrangements until they get the material with the properties that they seek. It is akin to the task of knowing how to start with piles of bricks, lumber, and shingles and end up with a house.

One level of this pursuit involves synthesizing materials of utmost purity, an ever more important requirement as engineers demand higher performance from materials. Optical fibers were high on the wish list even in the early 1960s, but every time anyone tried to make them, they had so many impurities that light would scatter, get absorbed, or otherwise get quashed into darkness before it could travel a city block. In the late 1970s, when fiber makers at Corning Glass Works nearly eradicated all impurities, they had opened the floodgates of optical communication. Purity of ingredients is also of paramount importance for semiconductor technologies, specialty alloys, solid-state lasers, and many other materials that are the stuff of advanced technologies.

Polymer makers have been developing several tactics for controlling how large molecules—each made of smaller molecular components linked into chains or networks—assemble into specific arrangements. Samuel Stupp, a materials researcher at the University of Illinois, for example, designs linear molecules so that they automatically fall into an arrangement like the teeth of a comb. Then he stitches them together more securely, using a measured dose of light that activates parts of each molecule to reach over and bond to its neighbor. What he gets from this procedure are so-called photonic materials that can manipulate light—amplify it or change its wavelength—the way electronic components manipulate electrons.[3]

Donald Tomalia, an entrepreneurial polymer chemist at the Michigan Molecular Institute, has been championing a new class of polymers called dendrimers because of their resemblance to the branching of trees. He and a small community of others around the world build one major kind of dendrimer starting with a small molecular core that they can link to a specific number of building blocks just like itself to form a slightly larger structure. Each of the newly linked components, in turn, can link to the same number of building blocks. By repeating this process a number of times, enormous yet identical polymer molecules branch out and build up, some up to the size of viruses. They are so uniform that Tomalia sometimes thinks of dendrimers as a class of enormous artificial atoms, which he hopes can be combined like real atoms into the dendrimer versions of molecules and materials.

Until recently, chemistry has been the art of crowd control so that the average behavior of huge numbers of atoms and molecules would result in the formation of materials needed to build and run society. Scientists are moving beyond that kind of crowd control and getting downright personal with atoms. They can now work with *this* or *that* atom, Joe or Jane atom, not merely with enormous numbers of atoms and molecules whose populations routinely could be denoted by numbers containing twenty-four zeros (that's a million billion billion).

The scanning tunneling microscope was originally designed for analysis, but its users will tell you that it doesn't take long to discover that the stylus can do things, real things, to the surfaces underneath. Scientists can routinely move individual atoms from place to place, using them to spell out words, university initials, even creating simple maps of the world in the ultimate expression of pointillist writing. They can pile up a few hundred gold atoms here and a similar number of copper atoms nearby so that a tiny flow of electrons goes between the piles. Behold—the world's smallest battery! They can use a laser beam to lay down fine chromium wires a mere sixty atoms wide to form diminutive wires for next generation electronic devices. They can create even smaller quantum dots whose electronic states can represents bits of data, dots so small that a stamp-size crystal riddled with such dots could hold the entire Library of Congress.

If thousands of scientists around the world weren't actually making such things, this all might legitimately be dismissed as the breathless fantasy of a technophile. But thousands of scientists around the world are making these things and showing off their atomic constructions in enormously magnified pictures projected onto huge screens at meetings or printing them in spectacular computer-hewn colors in their technical papers. Moreover, their skills are only becoming more sophisticated.

Phaedon Avouris, a physicist at an IBM flagship facility in Yorktown Heights, New York, knows this better than just about anyone. "With an STM you can image a surface on the nanometer or atomic scale, zoom in on a particular feature or region of interest, induce modifications on the atomic scale, and then image what you have done," says the Greek-born scientist, who grew up near Abdera, where Leucippus and Democritus first preached their atomism twenty-five hundred years ago.[4]

Researchers have sliced DNA molecules in half with an STM. They have created molecules by lining up atoms in the valleys of crystalline surface like eggs in an egg box.

Donald Eigler, at yet another IBM facility in San Jose, did this by lining up seven atoms of xenon, his favorite atom, on a crystal's surface. He scored an enormous PR coup for IBM in 1990, when his group wrote out the company's logo in xenon atoms. Many, many others also have used STM tips to draw lines, letters, and other symbols into crystal surfaces the way children draw in the sand. With an STM, though, the lines are drawn with atoms. Chips of the future may well be made with machines wielding large numbers of STM tips fashioning diminutive structures for storing data or computing answers to problems. (See Figures 15–20.)

Even before scanning probe microscopes were a gleam in Binnig's and Rohrer's eyes, Leo Esaki and Raphael Tsu, then also at IBM's Yorktown Heights facility, proposed another strategy for controlling matter on nearly atomic scales. If researchers could come up with a way of growing crystals that would enable them to mix and match elements known to form into semiconductor materials while controlling the spacings between the novel elemental combinations, they would be on their way to remarkable new materials. The physics that would go on within the new material structures, they reasoned, ought to make for faster electronic components or for devices that might interact with light in novel ways.[5]

One fantastic technique for doing this now, molecular beam epitaxy, is a kind of atomic spray painting onto surfaces, one atomic layer at a time. Every month researchers around the world publish their newest MBE constructions. In a stainless steel vacuum chamber of a typical MBE machine rests a crystal substrate on a temperature-controlled stage. Crucibles separated by shutters from the main chamber contain chemical or mineral precursors. When heated, these precursors emit atoms: gallium, arsenic, aluminum, indium, phosphide, or a

FIGURE 15

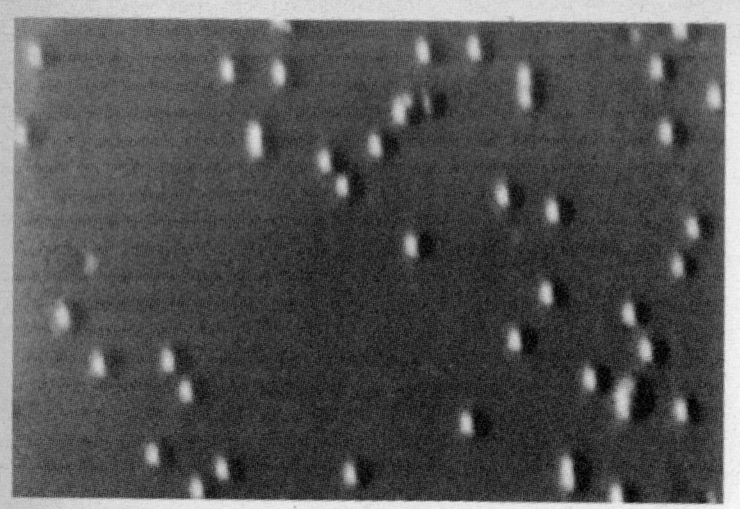

FIGURE 17

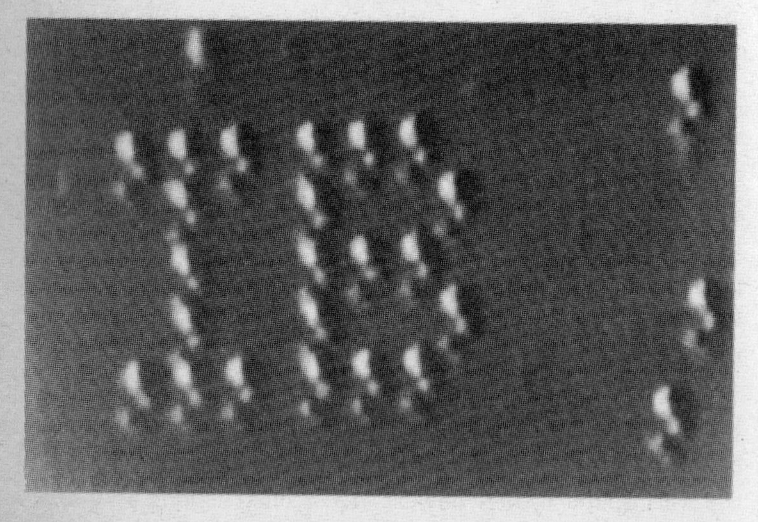

Writing with Atoms. In one of biggest PR coups of recent years, which also happened to be a technological tour de force, the physicist Donald Eigler of IBM's Almaden Research Center in San Jose, along with the visiting scientist Erhard K. Schweizer from the Fritz-Haber Institute in Berlin, used a scanning tunneling microscope (STM) to move individual xenon atoms on top of a surface of crystalline nickel to create the ultimate pointillist rendition of the

FIGURE 16

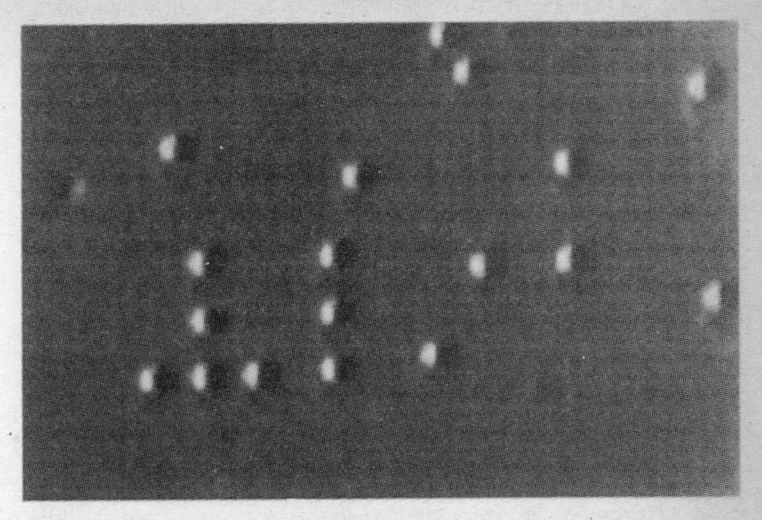

FIGURE 18

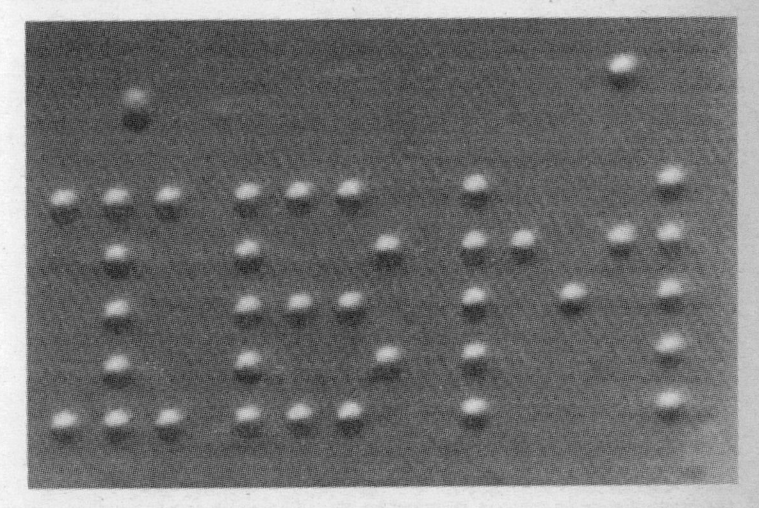

famous company's logo. The length of the logo is 660 billionths of an inch. This was one of the earliest demonstrations that tools like the STM held the technical promise of moving individual atoms at will. The sequence of four images shows how the researchers used the STM to transform an initial scattering of xenon atoms into a legible acronym, albeit at cryogenic temperatures and under a strong vacuum.

number of others. By using a computer to control how long the shutters are opened or closed, scientists can direct and lay down beams of different atoms on the substrate surface. Crystal growers can decide just how many layers of each element or elemental combination they want. The underlying crystal enforces a periodic structure on the incoming atoms like eggs taking their places on a huge egg carton into which they are being tossed. (See Figure 21.)

In the 1980s, as MBE became a more viable and user-friendly tool, researchers really began to see its possibilities: for example, it can synthesize solids whose electrons can take on energy states otherwise forbidden by the laws of quantum mechanics, absorb and emit new wavelengths of light, and interact with each other in different ways. Like painters getting hold of new colors, materials researchers have been building thousands of crystal structures with MBE machines, sometimes just to see what they might get.

More often now they have definite ideas of what they want. Maybe they seek new semiconductor materials in which electrons move faster so they can make faster chips to make faster computers. Maybe they want new solid-state laser crystals that emit in colors previously unavailable. (Red and green

---

On the other page are two more examples of atomic manipulation. The top image depicts the Kanji characters (used in the Japanese and Chinese languages) for the word "atom." These characters are written not in ink but rather in the medium of individual iron atoms precisely placed on a copper surface with a scanning tunneling microscope. Known as a quantum corral, the circular structure in the bottom image is made of forty-eight iron atoms moved into position on a copper surface by a scanning tunneling microscope. Twenty-thousand of these rings could be placed adjacent to one another across the cut edge of a human hair. This circular quantum corral, as well as other corral shapes, can confine electronic waves that undulate on the copper surface like tiny ocean waves; electron surface waves are about a trillion times smaller than ocean waves. Such atom-built structures could open routes to creating electron circuits that are smaller and more powerful than today's.

IBM CORPORATION, RESEARCH DIVISION, ALMADEN RESEARCH CENTER

FIGURE 19

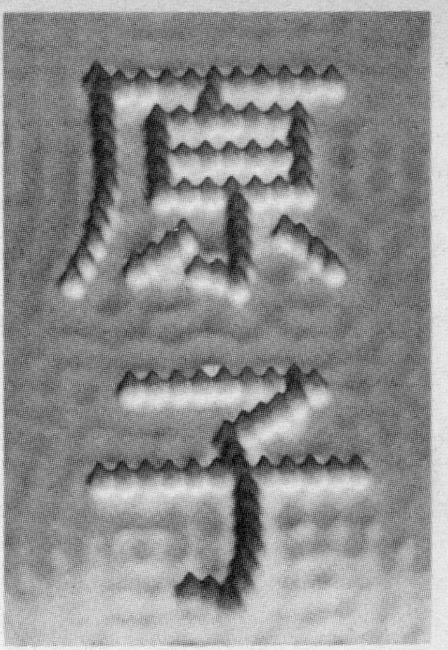

FIGURE 20

have been relatively easy, as is evidenced by the little red and green indicator lights on consumer electronics. Blue has been a tougher nut to crack, though in 1996 blue lasers made of such materials as gallium nitride had become the trophy for a technological horse race). In this way scientists and engineers have been able to create crystalline structures with specific quantum mechanical features not found elsewhere; in effect, they have found a way to get atoms and electrons to do things that were possible in theory but hitherto unrealized in practice.

One of the most visible and widespread successes of this approach so far is the lasers that read compact disks. The lasers are made of layers of such compounds as aluminum gallium arsenide and gallium arsenide arranged sandwich style, each layer on the order of tens of angstroms, a mere handful of atomic layers. The sandwich structure of these crystals creates a quantum mechanical context in which an input of electrons from the wall socket will cause electrons moving within the crystal to meet with positively charged vacancies in the crystal, or "holes." When electrons and holes meet, they can make light together.

Despite these extraordinary feats of atomic and electronic control, scientists still strive to achieve dominion over the coarser microstructural levels of materials as well. They still seek to control the way a material's ingredients—atoms, mole-

---

The MBE Medusa. It takes machines like this molecular beam epitaxy (MBE) machine to build materials one atomic layer at a time. Various ultrapure ingredients are placed into the ends of the stainless-steel appendages, where they are heated so that atoms or molecules bake off and then beam off toward a starting surface such as a crystal of silicon or gallium arsenide inside the MBE's vacuum chamber. By controlling shutters between the deposition chamber and the ends of the appendages, researchers can lay down virtually any sequence of atomic layers. With such tools researchers can construct new types of microchips, semiconductor lasers like the ones in CD players, and other solid-state gadgets.

BELL LABS/LUCENT TECHNOLOGIES

cules, powders, fibers, liquids, or whatever—aggregate during the heating, melting, solidifying, mixing, curing, and other processes that go into a material's formation.

Even in the netherworld between atoms and molecules on the one hand and perceivable pieces of material on the other, scientists are making headway. Besides MBE, other "deposition" techniques with equally formidable names—metal organic chemical vapor deposition, for example—have opened a vast set of possibilities for this layer-by-layer approach to synthesizing materials. Indeed, the catalog of synthetic techniques is as vast as the analytical catalog.

Chemists like George Whitesides at Harvard University design varieties of chemical building blocks that self-assemble into preconceived constructions.[6] Fraser Stoddart at the University of Birmingham, in England, has devised systems in which ring-shaped molecules slide over rodlike molecules, which are then plugged with molecular caps.[7] The ring can then occupy one of two sites along the rod, depending on whether it

FIGURE 21

is zapped with the light of one wavelength or the other. Stoddart hopes that these systems will provide the makings of molecular machines or logic components of a molecular computer. In enormous self-assembling arrays, these might comprise molecular memory storage devices so capacious that the entire Library of Congress could be carried around in one's pocket. The challenge is to find the right stuff for conveniently and accessibly stuffing ever larger amounts of data into ever smaller spaces.

Others are finding ways of controlling the way crystals nucleate and grow akin to the way living organisms grow and fashion the mineral hydroxyapatite into structures as intricate as skeletons. They are captivated by the idea of emulating nature, growing ceramic automobile engines in a manner that parallels the way a marine nautilus grows its beautiful spiral casing or a sea urchin its protective body of cathedral-like spines.

Among those in this field—called biomimetic materials research—is Mehmet Sarikaya of the University of Washington, who may have learned more from abalone shell-making than from his own professors.[8] To him, this pre-Paleolithic materials engineering on the part of the living kingdom is miraculous. Sarikaya and his colleagues need look no further than the nearest forest—or, for that matter, across the lab bench or dinner table for proof that atom-by-atom (or molecule-by-molecule) construction of the most sophisticated products known in the universe can proceed on a vast scale. Each cell of a tree or a human being is a tiny factory brimming with engineering know-how. It uses only readily available and inexpensive raw materials in the form of oxygen, hydrogen, or carbon dioxide from the air and from water; from food it extracts carbohydrates, proteins, or ions like calcium. Billions of years in the making, the cell's workshop of molecular machinery acts like a petroleum refinery to process that input into a huge variety of functional materials: stretchy and enduring vascular tissue, strong and tough tendons, hard and tough antlers and horns, muscle, bony skeletons, and compliant skins to support and house the entire operation.

The fantastic productivity of biological materials processing comes from several levels of multiplications. In each individual tree or person, trillions of cellular factories work in parallel. What's more, these living, breathing factories of biological materials replicate themselves as long as the raw materials are available and the instinct or will to do so remains. That amounts to a planet laden with biology's bountiful inventory of materials.

For biomimetic materials research nature serves as a perpetual in-your-face reminder of just how powerful human materials scientists could become. The living kingdom is a mass-production factory for a variety of astounding materials—structural, protective, mechanical, electrochemically driven substances that grow, strengthen, replicate, heal themselves, and adapt to their conditions, all within a limited, life-friendly range of physical conditions. Working in technological venues free of such natural constraints, materials scientists and engineers can envision processes of humanly devised materials production that would outstrip even nature's own genius and fecundity.[9]

Theory. *Synthesis* is the production of materials. *Analysis* breaks them down to reveal nature's secrets or to see what went right or wrong in the synthesis. *Theory* is both child and parent of synthesis and analysis; just as it is informed by the antecedent practical results of synthesis and analysis, so its concepts inform subsequent practical efforts that can steer researchers away from a lifetime of dead ends or into a lifelong adventure in fantastic material arenas.

In the first half of the twentieth century, the age-old trial-and-error approach in materials research began to transform with the coming of ever more sophisticated and detailed theories about all manners of materials. Ever vaster amounts of scientific observation were subsumed under conceptual frameworks that shed light on complex material phenomena such as the age hardening of aluminum alloys or the formation of diamonds from carbon atoms. Rather than relying on the deepest theories based on quantum mechanics—the early calculations of which were more often off than on—theories based on material defects, thermodynamics, and other higher-level concepts proved far more valuable than quantum approaches for explaining material behavior. And they often still do because they are better for depicting more humanly intuitive levels of material structure.

Quantum mechanical theory, which presumably harbors the deepest insights into materials, has remained limited as a practical tool by the computational complexity of its equations. After all, latent in Schrödinger's all-encompassing equation is the means of calculating any atom's or molecule's complete chemical personality: its shape, stability, and how and when it reacts under different conditions. Since materials are made of atoms and molecules, the same equation, in principle, harbors portrayals of material aggregations as big as bricks and sheets of steel.

But how to domesticate this wild mathematical behemoth? Paul Dirac, a contemporary of Schrödinger's, put it this way in 1929: "The underlying physical laws necessary for the mathematical theory of a large part of physics and the whole of chemistry are thus completely known . . . and the difficulty is only that the application of these laws leads to equations much too complicated to be soluble."[10] Dirac said so much nearly twenty years before the transistor was invented; though headway has been made, Schrödinger's equation continues to put up formidable armor against the computational cannons of the microelectronics revolution and its most powerful computers.

The reason why is that the exact solution describing the distribution of electrons around even a simple molecule such as ethanol, a globally popular nine-atom molecule, takes enough number crunching to give even today's supercomputers pause.[11] The simplest mathematical approximation requires twenty-six functions, one for each electron. That's just the beginning, however, because each electron and nucleus affects all of the others in an exponential web of interactions.

The calculation for ethanol molecules could end up involving more than five hundred thousand mathematical terms, each one containing a six-dimensional integral describing the physical features of the electron or nucleus. Moreover, solving these huge equations is not a single gargantuan one-pass computation like dividing a pair of million-digit numbers; rather it is a reiterative task in which the results of each earlier pass feed into the next one and so on for tens of thousands or more cycles, or time steps. Since each cycle represents a billionth or a millionth of a second of activity in the life of the molecule, it takes that many before a good idea of the molecule's actual behavior becomes apparent.

Scientists mostly have surrendered to the complexity of the calculations by coming up with simplifying approximations. These have been more or less artful exercises in self-imposed blinding. The strategy is to ignore big hunks of the quantum mechanical equations that ultimately have little to do with the actual chemical behavior of atoms and molecules. Why worry much about the six inner electrons of an oxygen atom—and the hundreds or thousands of additional mathematical terms needed to describe them—if it is the atom's outer two electrons that are responsible for virtually all of its chemical behavior?

That strategy of selective editing was a good idea in principle. Yet any approximation that stood a chance of being solved on the earlier generations of computers was likely to yield calculations and predictions that were either dead wrong or subject to an unacceptably wide range of uncertainty or error.

In the mid-1980s there were only a couple of dozen examples of calculations that had predicted new compounds and reactions.

Most calculations simply made sense of what experimentalists already had known and observed. By the mid-1990s the number had grown to hundreds and appeared on an exponential trajectory reaching into fields as diverse as rocket fuel development, steel alloy design, drug design, and semiconductor engineering. "We now can become the world's expert on a particular molecule or material even though we have never seen it," remarks Marvin Cohen, one of the most outspoken champions and practitioners of computational materials research at the University of California at Berkeley.[12] As early as 1970, he had predicted successfully that silicon, a semiconductor at room temperature and pressure, would behave as a superconductor under certain extreme conditions.

Cohen and his former graduate student Amy Y. Liu at the University of Berkeley made headlines in the summer of 1989, when they reported calculations for a hypothetical material made out of carbon and nitrogen in a ratio of three to four. They predicted that the material might even be harder than diamond, the hardest substance yet known in the universe.[13]

They calculated many possible crystalline arrangements of the carbon and nitrogen atoms. They then used other theories relating atomic structure to hardness to infer that some of the resulting in-silico structures ought to represent materials that are at least as hard as diamond, and maybe harder.

Since this prediction, various researchers have intermittently reported evidence of making carbon-nitrogen materials nearly as hard as diamond. Some have gone to extremes: squeezing the ingredients at great pressure between the jaws of anvils made of diamond or using ion guns to propel nitrogen atoms into a carbon-rich surface in the hope that the sought-after carbon-nitrogen structures would arise spontaneously. Although researchers have reported signs of materials that seem as hard as diamond, their samples have been too tiny to make confident assessments. And other theorists have been questioning Cohen's and Liu's original calculations, coming up instead with alternative crystal arrangements that might yield the elusive harder-than-diamond stuff.

There are other bridges that link the mind-numbing quantum mechanical equations with older and more intuitively grasped theories about how matter and materials behave. Many of the estimated two thousand computational chemists and materials scientists spend their days developing computational strategies and algorithms that serve as those bridges.[14] Often they will ease up on the aesthetic appeal of pure prediction by incorporating experimentally derived data that can help computers direct their computational efforts more quickly toward sensible. With that strategy solid-state scientists have been extending computational chemistry way beyond small discrete molecules like ethanol to the architectures of periodic materials such as crystals with atomic constituencies numbering in the millions.

In addition to using empirical data, computationally inclined scientists adapt Newtonian ideas to molecular realms, linking classical physics to thermodynamic accounts of how the subtle redistributions of energy throughout a system of atoms and molecules will lead to this or that crystal or less-ordered amorphous phase, grains of this or that size, and other kinds of microstructural details. These so-called molecular mechanics techniques can sidestep computer-gagging quantum mechanical calculations by portraying molecules as, say, sets of balls linked by springs whose tension and bounce represent the balance of electron-electron repulsions and electron-nucleus attractions. Or they might rely more on matching the energies of particular regions of a mix with "phase diagrams" that codify the kind of microanatomy that can form under those conditions. These types of computations, of course, often can first be anchored to quantum mechanical calculations to make sure they can be grounded in first principles.

Just as the theories of molecular mechanics theories can link to those of quantum mechanics, so too can they link to well-developed theories of thermodynamics and kinetics that account for bulk properties of materials: when they melt, crystallize, or vaporize. These higher-level theories, in turn, link to engineering practice, through which real materials become useful things.

Computational materials science is getting reliable enough that researchers have even begun to think of "in-computero" activity as a new category of scientific activity somewhere between theory and experiment. Consider a computational chemistry group at the National Aeronautics and Space Administration's Ames Research Center in Mountain View, California, where computational studies on hydrogen atoms recombining into hydrogen molecules were funneling into engine design work on the on-again/off-again National Aerospace Plane. The NASP (which has had several aliases) is envisioned as an aerospace craft that takes off and lands like a jet but can go into and out of orbit as well. Designers have used computational science to help determine optimum lengths for nozzles and combustion chambers and to establish more precisely the kinds of hostile conditions that materials of these engine components will have to endure. These reactions occur so quickly in such hostile arenas that only the scantiest of clearly relevant experimental data from traditional measurements are possible anyway.

"In the last few years, companies specializing in polymers, ceramics, and materials in general have come to believe that theory can now provide good enough information that they can do fewer experiments," reiterates William A. Goddard III, director of Materials and Molecular Simulations Center at the Beckman Institute of the California Institute of Technology.[15]

As computers grow more powerful, computationalists grow more adept at exploiting their scientific potential. There are visionaries who are toying with novel approaches that seem to border on the fantastic. Take the creative computer architecture of MIT researchers Tomasso Toffoli and Norman Margolus: rather than building computers out of an enormous number of on/off switches—gates that let electrons through or block their flow—Toffoli and Margolus have been building what they call "cellular automata machines," or CAMs. The basic unit of a CAM is a "cell" whose own state depends on the collective state of its neighbor cells. Programmers stipulate some rule by which the states of neighbor cells determine the state of each center cell. The computer then changes the cells' states accordingly, and the procedure is repeated until a steady state or perpetual static or something in between emerges from the collective.[16] This kind of computer architecture—with discrete cells interacting with neighbors—resembles the way atoms or molecules behave in collectives. Tomasso, when he wears a materials research hat, thinks of his computer as "computronium," a universal chemical element. In his mind, computronium is a computational chameleon capable of simulating any material thing or process such as crystallization or alloy formation with far more physical verisimilitude than simulations run on more conventional machines.

The triumvirate of analysis, synthesis, and theory has set off an almost dizzying cycle of positive feedback among materials researchers over the past thirty or forty years; their interaction has inspired fresh, diverse efforts ranging from research into what makes the sea urchin's spine so hard to nothing less than solving major social problems such as energy-efficient forms of production, pollution, and transportation. When viewed collectively, the materials research community has become something like a vast search team, turning over every stone and leaf, looking up every tree, and scouring every attic to see what interesting or useful trinkets of matter can help to change the world.

FIVE

the MATERIALS SERENGETI

Even though materials science and engineering has been a bona fide field of research for only a few decades, it has spawned a diversity of products, projects, and practitioners rich enough to populate its own version of the Serengeti National Park in Tanzania.

One of the very best times to go on safari is just after Thanksgiving each year. That is when about four thousand materials scientists and fellow-travelers converge on Copley Place in downtown Boston for the annual winter meeting of the Materials Research Society (MRS). They come from every state in the union, Tokyo, Kyoto, Seoul, Moscow, Berlin, Paris, Manchester, Zurich, Johannesburg, Bombay, Paris, and hundreds of other places. They come from universities, government labs, and industrial research facilities. High-tech venture capital firms send representatives. Hundreds of companies eager to do some business set up booths and equipment in a massive expo complete with bowls of chocolate and attractive salespeople to lure potential buyers of mass spectrometers, nanoscopic ceramic powders, and the hottest new scanning tunneling microscope. This town-size conclave fills entire hotels, overruns restaurants, and injects millions of dollars into Boston's economy.

Mostly, these masses come for a materials research show-and-tell nonpareil. They come to talk science, reunite with graduate school pals, gather intelligence about competitors, and to talk with long-distance collaborators. They come to buy and sell instruments, find jobs, and locate new sources of alumina powders for making refractory bricks or maybe isotopically purified methane for synthetic diamonds that are more perfect than any sparkling gem. Everything you always wanted to know about materials research and a thousandfold more can be learned during these weeklong pilgrimages. An MRS meeting is like a carnival for the pocket-protector set.

Because advances in materials have had everything to do with general scientific progress no matter what the field, this fair is a general celebration of all that science and engineering have wrought: the history of microscopy is partly a history of glassmaking; the history of spectroscopy is partly a history of light-sensitive materials for the detectors; every computer-controlled instrument in every laboratory owes a debt to semiconductor technology, which is as materials-centric as you can get.

When MRS convention participants pay their registration fee, they get a telephone book of abstracts describing several thousand talks. At any given hour during the meeting, twenty-five different researchers will be talking simultaneously to audiences in twenty-five different meeting rooms. The technical torrent rages all week. Specialized thematic symposia run through the meeting like subplots of a play. The symposia become havens at which each participant can find his or her own familiar ground.

There are polymer specialists and synthetic diamond aficionados. There are users of molecular beam epitaxy machines who create new kinds of transistors and lasers on a chip by building up special crystals one atomic layer at a time. There are solid-state physicists placing bets that crystals with "giant magnetoresistance" (materials whose electrical resistance changes sharply in the presence or absence of a magnetic field) are the key to next generation data storage and handling technologies. There is a perennial surfeit of researchers studying a class of ceramic superconductors that can carry electricity resistance-free at higher temperatures than all previously discovered superconductors.

There also are theoreticians sharing ways of coaxing more out of Schrödinger's equation or of Newtonian descriptions of molecular activity. They work in odorless laboratories awash in the glow of computer screens. Their equations are the wands with which they conjure up an endless vista of new materials that will lead to a technological landscape as far from today's as today's is from that of the early century.

There are researchers at MRS meetings who rhapsodize in minute molecular detail about designing new "energetic" materials, solids, and liquids that are good as propellants or explosives that pack ever more bang or thrust into smaller amounts (mind you, all of this has to be possible with minimal danger of accidentally exploding the stuff during preparation and handling). You won't hear the most secret stuff about energetic materials since the doors of an MRS meeting are open to all registrants regardless of their security classifications. But there are other meetings organized by organizations with names like the National Defense Preparedness Association where participants need security clearances and the doors are guarded.

Still other research tribes at an MRS meeting obsess about safe ways of perpetually entombing radioactive waste into glassy materials that can take the bombardment of radioactive slings and arrows for millennia. Another group aims to emulate the arachnidian silken feats of natural polymer engineering and use them to make what they call "biomimetic materials." For this tribe the exoskeleton of a bess beetle could well provide the key to a new generation of aircraft: lighter-weight, more resistant to damage, and harder to detect.

Yet amidst the apparent mishmash of tribal differences lies that unity of thought and action that Northwestern University's Morris Fine had seen even in his Bell Lab Days in the 1940s. It is the same conceptual unity that provided the very impetus for the birth of the original Interdisciplinary Labs (IDLs) at Northwestern, Cornell, and the University of Pennsylvania and for the founding of the Materials Research Society itself.

As Fine pointed out years ago, material research is not just chemistry or quantum theory or solid-state physics or process engineering or marketing. It is all of these as well as any other way of thinking that helps reveal how the structure, properties, performance, and processing of materials relate to one another, to technology, and societal needs. The founders of MRS knew that the IDLs were merely a start. The age-old disciplinary segregations would not buckle easily even with new academic structures such as the IDLs in place. The founders of the MRS wanted to create another kind of forum beyond IDLs at particular universities; they sought a more general infrastructure for changing attitudes, a task as daunting as any technical challenge.

Materials researchers have developed a collective insight into the ways of materials that often even astounds themselves. Their scientific conferences have become show-and-tell marathons in which researcher after researcher, in darkened hotel meeting halls, flashes annotated series of slides depicting exquisitely detailed data and images about a material's structure, functions, and place in the technological world. Many of these slide shows end up winning the attention of only a handful of like-minded specialists. Most quietly recede into oblivion. Some, however, take root and even grow into great branching subdisciplines. And some of these trigger research wildfires that change the world.

Going beyond this impressionistic helicopter view of the Serengeti of materials science and engineering means making forays on foot amidst the different tribes, prides, and packs. You could probably mingle with a different group for every day of the year, but a sampling will suffice to show the diversity on the plain. Welcome to the tribes of synthetic diamond, biomimetic materials, and smart materials and structures.

SYNTHETIC DIAMOND

The dream to make synthetic diamond spans over a century. Some of the nineteenth century's most decorated scientists went to great extremes to coerce carbon-bearing ingredients into the coveted crystalline structure of diamond. In 1880, for example, the Scottish chemist James Ballantine Hannay filled tubes with various carbon-containing ingredients such as burned sugar and bone ash, sealed the tubes, and then heated them until they exploded. He claimed that tiny hard particles retrieved from the debris were diamonds. (He also attributed a nervous disorder he developed to the frequent fear of injury during these explosions.) Later analysis of his experimental protocol virtually precluded the possibility of his actually having produced diamonds. Robert Hazen, a geophysicist at the Carnegie Geophysical Laboratory in Washington, D.C., speculates that Hannay's assistants may have spiked his preparations with tiny diamonds with the hope of getting Hannay to give up his mortally dangerous obsession.[1]

The first documented successes occurred in 1953 and 1954. The earliest one was quiet and remained unannounced by cautious researchers at Sweden's main electrical company. The other was announced with fanfare in February of 1955 by the General Electric Company before the Department of Defense silenced the company for several years in the interest of national security. The youngest member of the GE diamond-making teams, the freshly graduated Robert Wentorf, Jr., even found that he could convert peanut butter, a rather popular carbon-containing material in its own right, into diamond grit.[2] At the time, industrial diamond was considered a strategically important material since it was used to make tungsten carbide machine tools. Tungsten carbide machine tools could cut and mill steel into munitions, tanks, frames, and other military materiel. Machinists would have loved to use diamond itself, which is much harder than the carbide tools. Unfortunately, carbon—even in the form of diamond—dissolves in iron and several other metals at elevated temperatures, thus precluding its use for high-speed machining of iron-based metallic alloys including steels. Moreover, the superlative hardness of diamond means it is very difficult to shape it into something like a drill bit or milling blade.

The early successes at GE depended on a punishing heat-and-press process designed after nature's own presumed way of making diamonds deep underground in magmatic cauldrons. Thousand-ton presses, themselves wonderful feats of engineering and materials development, and temperatures of 2,500°F (and a catalyst such as a piece of nickel) hot-squeezed graphite into diamond grit. Under these processing conditions, the graphite's carbon atoms dissolved into the catalyst and then crystallized in the form of tiny diamonds like salt crystallizing from a saltwater solution.

This transformation on the atomic level is as basic as it gets. Each carbon atom in graphite is bonded to three other carbon atoms in a flat arrangement. In a piece of graphite, this three-way bonding arrangement goes on for billions and billions of atoms ahead, behind, to the right, and to the left.

Carbon Jungle Gyms and Chicken Wire. The differences between diamond and graphite goes atomically deep. Each carbon atom in a diamond is strongly bound to four neighbor carbon atoms to create an everywhere-the-same crystal. This leads to a symmetric crystalline network of famously superlative strength. Each carbon atom in a piece of graphite is bound to only three other carbon atoms, leading to flat, chickenwire-like sheets of carbon that then loosely stack upon one another. The resulting soft, dark material is the stuff of pencil "leads" and lubricants.

Each resulting sheet is like an endless expanse of molecular chicken wire. Above and below each sheet of molecular chicken wire are other similar sheets. The overall structure, therefore, resembles an enormous stack of sheets of molecular chicken wire. When you drag a graphite pencil on a piece of paper, little sheets of molecular chicken wire get left behind. (See Figure 22.)

Diamond, on the other hand, is made of an everywhere-the-same structure. Every carbon atom reaches out and bonds tightly and strongly to four neighbors in a tetrahedral geometry. Every angle and distance between carbon atoms is the same. Making synthetic diamond from graphite entails the formation of many sheets of molecular chicken wire into a single infinite molecular jungle gym. Every carbon must bond to four others, not only three. Loosely bound sheets of carbon must rearrange into a single monolithic structure in which there are no distinct layers. Diamond's superlative hardness, compressibility, and heat conductivity come from a combination of the crystal's supreme symmetry and from the strength of the carbon-carbon bonds inside.

FIGURE 22

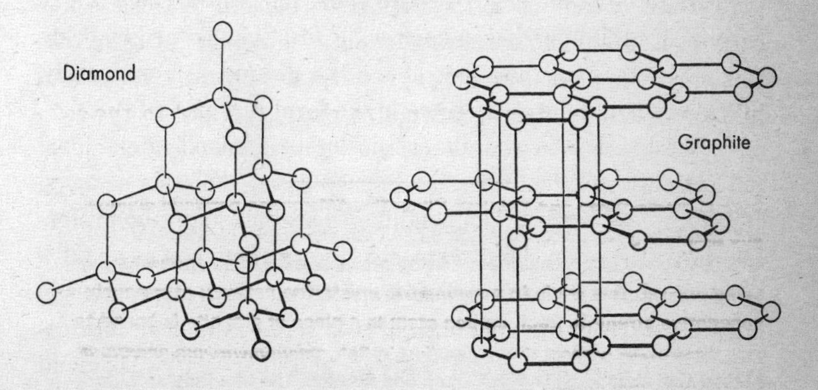

GE filed patents for their heat-and-press process for making synthetic diamond in 1959, just days before DeBeers Consolidated Mines had filed in South Africa. DeBeers had already been a monopolizer of gem quality and industrial diamond for decades. Despite Herculean legal efforts to prevent it, however, GE broke DeBeers's lucrative monopoly of the industrial diamond market. The diamond grit industry has since grown. GE still dominates, but there are other major players, including DeBeers, which bit the bullet and licensed GE's technology. In the 1990s the industrial synthetic diamond market has been approaching the billion-dollar mark.

Diamond grit has a limited range of applications, almost all of them based on diamond's hardness and abrasive properties. If you could make diamond in larger formats like flat sheets or films, however, the material's other material talents might find outlets. The technological wish list is a long one. And business analysts predict that the market for diamond films would make the diamond grit business seem like peanuts.

Since diamond is the hardest substance known, transparent diamond coatings would perpetually protect underlying surfaces—finally, a truly scratch-resistant eyeglass lens, for example. A diamond film on a computer's fast-spinning hard disc would protect it from data-destroying crashes by the read/write head hovering a mere thousandth of an inch above the rapidly spinning disc, a harrowing situation likened by some to a 747 screaming just feet above the ground. Unfortunately, diamonds do not normally come in the form of films.

Because it absorbs almost no light, diamond offers ideal protection for optical fibers and space-based radiation sensors. But here, less perfect "diamondlike" materials might be preferable since they deposit as a smooth layer rather than as a multifaceted, light-scattering landscape. For the same reason engineers expect diamondlike coats to serve as lifelong lubrication for such things as ball bearings and gear shafts. Diamond could spell the end of a lot of the wear and tear that costs billions of dollars in automobile and machinery repairs.

Diamond carries heat away at rates second to none, almost never reacts with anything (and then only at high temperatures), and is an excellent electrical insulator. Its chemical inertness and nonstickiness make it a prime candidate for surgical blades and implants such as artificial hips. Some scientists even envision semiconductive forms of diamond that might supersede silicon and gallium arsenide as the material basis for a yet faster generation of electronic chips.

"Diamond would be the best semiconductor if you could make it work," says Rustum Roy of Pennsylvania State University.[3] Roy is a founding member of the MRS and a regular on the blue-ribbon panels that filter through government agencies. He was the first to convince government research managers at the Office of Naval Research to fund work on a way of making synthetic diamond into the flat and extended forms required for such applications as diamond electronics.

The first hint of Roy's diamond vision came into view over fifty years ago, just a year after the first semiconductor transistors were reported and several years before the General Electric researchers had developed the high-temperature, high-pressure root.

The demiurge for low-pressure synthetic diamonds was William G. Eversole of the Union Carbide Corporation (which got its start in 1898 making calcium carbide for lighting). He began his diamond quest in his company lab in 1949. He scored his first successes in the period between November 26, 1952, and January 7, 1953.[4]

Eversole's method involved heating small diamond seed crystals contained at low pressures in a sealed vessel heated to at least 600°C. Meanwhile, he let methane or carbon monoxide molecules decompose inside of the heated vessel. Liberated carbon-containing fragments from the methane molecules settled onto the seed crystals, slowly and subtly enlarging them. The process is now known as chemical vapor deposition (CVD). The project-killing rub was that the growth rates were so slow that it would take many months of continuous deposition before a film as thick as this page would grow.

Eversole was swimming against the theoretical current of the time, which held that diamond crystals would fail to start growing under low-pressure conditions where diamond is the unstable form of carbon and graphite the stable form. But Eversole knew that organic chemists often succeed in making "unlikely" chemicals. The trick is to isolate certain so-called metastable chemical structures before they have a chance to rearrange into more stable molecular arrangements. Samurai sword makers were relying on the same principle when they plunged their hot blades into their water baths. That way they locked the mix of carbon and iron atoms in such a way that the microstructure yielded a steel edge with famous strength and toughness. Though Eversole succeeded in using diamond seeds to engender the growth of new layers of metastable diamond, the sluggish growth rate spelled doom for the method's commercial prospects. The project fizzled out.

For years afterward the party line among diamond-makers was that CVD wasn't commercially viable, a perception that was strengthened by GE's 1954 patent on the high-temperature, high-pressure technique. It made more sense. To make their diamond grains, they subjected graphite to the enormous pressures and searing temperatures typical of 240 or so miles below the earth's surface, where natural diamonds form. Under such extreme coercion, the carbon atoms in a graphitic arrangement reshuffle into diamond crystal, the stable form of carbon. That is how you get graphite to become diamond. Right? Right.

Not until the early 1960s did Eversole's work appear in the patent literature (it never appeared in a peer-reviewed technical journal). That was just about when John Angus got started in the field. Angus, now a veteran chemical engineer at Case Western Reserve University in Cleveland, was told by his colleagues that he was crazy to think anyone could make diamond by letting heated mists of carbon-containing chemicals deposit onto other surfaces within a vacuum chamber.[5]

But Angus and a handful of others knew too much to dismiss CVD. If they could get it to work, CVD diamond-making would be a milder, less apocalyptic alternative to the GE method. Still, most would-be CVD diamond growers abandoned the dream with one look at carbon's phase diagram. (A phase diagram is a graph that maps out the gaseous, liquid, and solid forms [or phases] a specific material assumes under various combinations of pressure and temperature. Phase diagrams were born in the nineteenth century, when physicists were first learning the rules of thermodynamics, which govern the way heat, other forms of energy, and materials interact.) Carbon's phase diagram suggests that carbon atoms ought to assume graphitic chicken-wire arrangements under the relatively low pressures of CVD.

The CVD diamond diehards like Angus reminded themselves, however, that a phase diagram, like any map, is an abstraction. Under certain conditions then yet to be discovered, the unstable form of carbon—diamond—might freeze out before it had time to settle into its preferred and more stable form—graphite. If Eversole had found conditions suitable for painfully slow growth, perhaps someone else would find faster-growing conditions.

"You really only have to know a few concepts about how crystals are put together and then make some intuitive guesses," remarked Robert DeVries, a consultant to companies that make hard materials and a retired GE diamond-research veteran. "Most of these advances are made by intuitive materials scientists playing around in a sandbox."[6]

Carbon-containing gases like methane—often maligned as

"swamp gas"—and ethylene represent CVD diamond-making "sand." A vacuum chamber maintained at pressures ranging from one atmosphere to less than a thousandth as much serves as the "box." Inside this box a source of molecule-busting energy—such as heat from a red-hot tungsten wire or from microwaves—decomposes the carbon-bearing molecules carried into the chamber on a steady flow of hydrogen gas. Carbon fragments ripped from these molecules voraciously search for new places to bind. They find what they are looking for on target surfaces also in the chamber. That is where the carbon ends up depositing as a diamond coating. Carbon atom by carbon atom, diamond films form and slowly thicken on these surfaces.

Much of the early CVD diamond research proceeded by trial and error: feed in a little more or less gas, add or subtract hydrogen, raise the substrate temperature a little. Playing in sandboxes can be a lot of work, however. Not until 1968 did Angus successfully repeat Eversole's experiments. In fact, he went further, more than quadrupling the growth rate. He found that swamping the carbon-containing gases going into the CVD chamber with hydrogen gas seemed to hinder or perhaps reverse graphitic deposits that would otherwise cover the surfaces of the diamond seeds and prevent further growth. Nothing much came of this achievement, at least not immediately.

Three years later, however, Angus described the role of atomic hydrogen (as opposed to the two-atom molecular form of hydrogen) at a meeting in Kiev. One of the participants was the Russian scientist Boris Deryagin, of the Institute of Physical Chemistry in Moscow. Deryagin later become a pivotal player in the development of the CVD diamond game. Angus's work was the springboard that propelled Deryagin and his colleagues to the forefront of the low-pressure synthetic diamond race.

The Moscow researchers actually had begun their own CVD work by 1956, using molecules such as CBr_4 and CI_4, which would shed their bromine and iodine atoms under conditions of 1000°C and a few billionths of an atmosphere, a tough pressure to maintain. They saw far more promise in Eversole's technically simpler route, which required low but more reasonable pressures. By the end of 1969, the Russians, too, had confirmed Eversole's work. And they had throttled up the growth rate more than tenfold, a rate that would have begun turning industrial heads had convincing word of their success gotten out.

Both the Angus and Deryagin groups knew that hydrogen seemed to increase growth rates of the diamond films by frustrating graphite growth more than diamond growth. But eventually graphite islands would still develop on the diamond surfaces, frustrating further growth. Only by first cleaning these islands away with reactive hydrogen or oxygen atoms could the researchers continue diamond growth. Repeating this sequence yielded more diamond growth. But this stop-and-go method was still too slow and awkward to be practical.

Throughout the 1970s Deryagin and his Russian colleagues were learning in earnest the critical importance of atomic hydrogen. They found that the presence of atomic hydrogen during the growth of the diamond crystals thwarted graphite islands from nucleating on the diamond seed surfaces by temporarily binding to those surfaces and physically blocking graphitic carbon-carbon bonds from forming. And if graphite did form, the hydrogen atoms could break it up by swooping in and snapping up carbon atoms like hawks grabbing mice. Deryagin's group was putting its growing basic understanding to use to find conditions that would optimize diamond growth.

Most materials scientists were sleepwalking around these developments. Besides Angus and a handful of others, no one picked up on the implications of the Soviet work, partly because of what the Soviet diamond-growers had left out of their papers. Their technical reports on the subject were conspicuously devoid of quantitative details about the pressures and temperatures they used. They didn't reveal enough for anybody to reproduce the work. For other CVD researchers, that lack of detail initially rendered the Russian papers more skimming material than study material.

On top of that, Deryagin's group had all but killed its credibility. In the late sixties and early seventies they became leading players in the infamous polywater episode, a pseudoscience fiasco now often recalled as a precedent for the more recent cold fusion saga. For a time researchers thought they had discovered polymeric forms of water and that in principle the entire ocean could become one giant polymer. As a result of its role in the polywater affair, an aura of charlatanry surrounded Deryagin's group.[7]

Nevertheless, Japanese researchers, particularly adept at gleaning nuggets of practical value from the world's scientific archives, took the Russian work seriously. A team led by Nobuo Setaka at the National Institute for Research on Inorganic Materials (NIRIM) made its first diamond films in 1974 by passing a methane-hydrogen mixture by a hot metal filament in a chamber kept at low pressures. Not much new there. But the Japanese crew was discovering specific conditions that seemed more promising than those attained in any earlier attempts.

After seven more years of systematic efforts to optimize the rate of diamond growth, they reported their increasingly successful work to a gathering of Japan's Carbon Society. And in a head-turning spate of developments between 1982 and

---

Better-than-Nature-Diamond. Synthetic diamond materials like the one seen behind the microelectronic circuit above are made by taking carbon-bearing starting materials such as graphite or methane gas and rearranging the carbon atoms in these ingredients into pure diamond. Nothing conducts heat better than diamond, which makes diamond particularly good for shunting destructive heat away from hard-to-reach or particularly critical chips. The synthetic heat sink above conducts heat more efficiently than natural diamond because researchers made it using a carbon source enriched in one of the several forms (isotopes) of carbon atoms. In natural sources carbon exists as a mix of these several atomic forms and natural diamonds reflect that mix.

GENERAL ELECTRIC RESEARCH AND DEVELOPMENT CENTER

1984, they showed the world how to make diamond films with growth rates of at least several microns per hour. The magnitude of this rate, thousands of times faster than Eversole's and hundreds of times faster that Angus's or Deryagin's, stunned the research community. No one had even come close.

"The current worldwide interest in new diamond technology

FIGURE 23

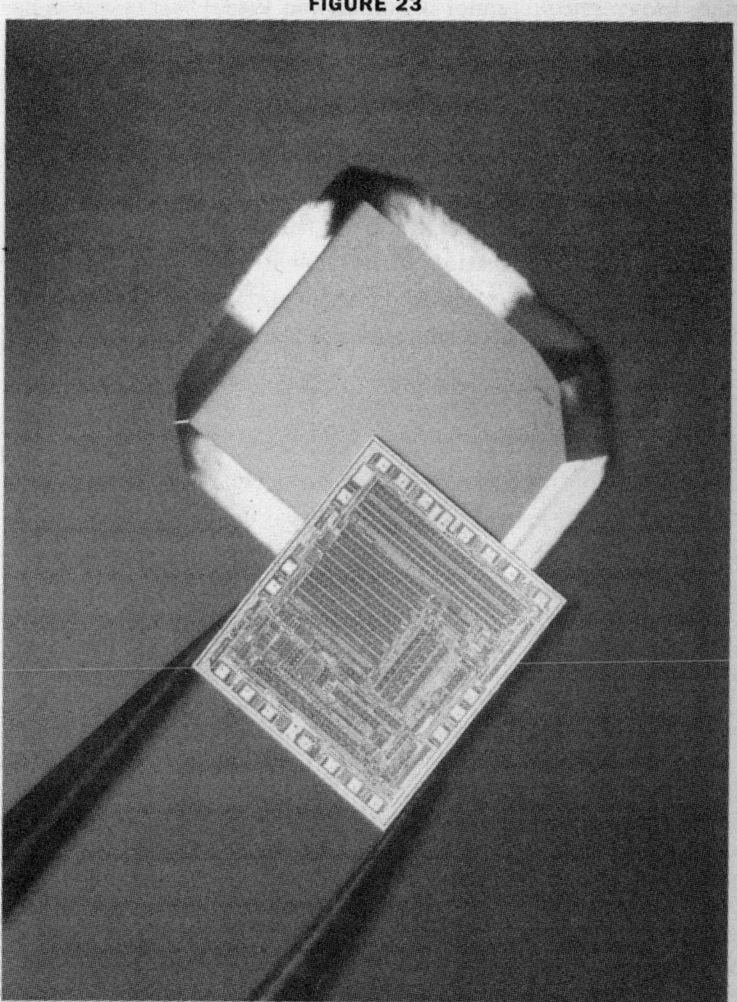

can be directly traced to the NIRIM effort," says Angus.[8] As if to remind the world of this remarkable alchemy, one Japanese researcher even reportedly made CVD diamond film from sake, the Japanese rice wine. The carbon source in this case must have been carbon-bearing molecules such as ethyl alcohol. It was CVD diamond's answer to Wentorf's high-pressure, high-temperature peanut-butter stunt. (See Figure 23.)

Ever since its quiet beginnings in 1949, CVD research has been taking the Black Art out of diamond-making by deciphering the detailed chemical and physical mechanisms underlying metastable diamond growth. In the late eighties and into the nineties, this search has become an international technology horse race, with Japan leading in commercializing the nonoptimized technology and the United States having the edge in basic diamond deposition science. That latter edge in fundamental science could harbor the most important keys to unleashing diamond's technological and commercial promise.

Still, many of the physical and chemical details underlying the CVD process remain mysterious. Just what types of molecular fragments break from the carbon-containing gas molecules? How do they nucleate into tiny diamond seeds and then aggregate into larger diamonds? These unanswered questions have not prevented researchers from discovering reliable conditions the old-fashioned way—trial and error, lubricated by an admixture of intuition, experience, and theoretical insight.

By regulating the hydrogen-hydrocarbon ratio, the pressure inside the chamber, the temperature of the target materials, and the means of ripping up the carbon-bearing precursor molecules, researchers have grown diamond films at up to one thousand microns per hour—about a penny-width—onto relatively large and intricate target areas. Like crowded dollops of cookie batter baking on a baking sheet, thousands of tiny flat diamonds nucleate on the surface and then grow outward into a continuous polycrystalline film. One elusive goal, which could advance diamond beyond silicon and gallium arsenide as the ultimate semiconducting material, is to grow a film consisting of a single flat diamond crystal; unfortunately, the irregular and hard-to-control boundaries between crystallites in a polycrystalline sheet run counter to the uniformity needed in electronic materials such as silicon.

From their pole position in the practical art of diamond film growing, the Japanese CVD diamond community has been in a full sprint. This is no surprise to Rustum Roy, a vocal champion of CVD diamond. He remembers visiting the lab of one of his former Japanese graduate students in 1984. "When he showed me the diamond films he was making, I couldn't believe it," Roy recalls.[9] Roy convinced the U.S. Office of Naval Research to begin funding basic CVD diamond research lest the United States fall hopelessly behind in this critical technology race. That marked the awakening of the federal government to the promise of diamond films.

Two years later Roy and a colleague organized the country's first diamond film technology consortium involving researchers at Penn State and more than thirty corporations betting that diamond films could become big business.

In the summer of 1990, the National Research Council, an arm of the prestigious National Academy of Sciences, which advises the federal government about scientific and technological issues, publicly placed bets on CVD diamond in a report, *Status and Applications of Diamond and Diamond-Like Materials: An Emerging Technology.* "The ultimate economic impact of this technology may well outstrip that of high-temperature superconductors," the NRC panel suggested.[10] Business analysts have made predictions that a low-pressure CVD diamond industry could haul in billions in the coming years.

None of this is a surprise to Angus, who was a pioneer and proponent of CVD almost twenty years before Soviet and Japanese scientists finally showed the world the practicability of the method. Angus is too humble to say "I told you so," but he could say it. What he has said is that "diamonds are going to be everywhere. . . . They'll be in pots and pans, on drill bits, and razor blades, in copying machines and on hard disks."[11]

Given the diversity of the synthetic diamond research community, his prediction seems a good bet. Like the potters of an earlier era, researchers around the world have developed variations on the basic CVD diamond-growing techniques. Some are better suited for rapid deposition. Other techniques are slower but produce films of finer quality. Still others don't yield true diamond films but rather diamondlike films. The smoother texture of these films could make them attractive for applications in which the tiny sharp facets of diamond films would cause wear or light scattering: coating ball bearings or telescope mirrors or radomes of smart missiles. By scaling up the experimental CVD apparatus in today's labs, scientists could usher in an era in which products made with diamond and diamondlike film could become as common as plastic products are today. (See Figure 24.)

One especially bold group at Lawrence Livermore National Laboratory once even proposed to make eighteen million carats worth of CVD diamond as the basis for the world's most sensitive particle detector, which they claimed would serve well inside of the ill-fated superconducting supercollider.[12] They argued that nothing could surpass synthetic diamond in both its ability to withstand the bombardment of high-energy particles and the sensitivity with which it detects the impacts of the particles. The sensitivity is due to the fact that electrons move far more quickly in diamond than they do in silicon. The

---

Diamond Fields. Carbon-bearing molecules were broken apart inside a chemical vapor deposition chamber, and the liberated carbon atoms then rearranged into the diamond film revealed here by a scanning electron microscope. Diamond films like this can be deposited onto many different surfaces, even ones with intricate shapes. The region framed off on the left appears at higher magnification on the right, an area that would just about fit atop the cut edge of a human hair.

THE NORTON COMPANY

proposal never got the thumbs-up, a rejection that became moot anyway when Congress killed the supercollider project in 1993. That anyone could seriously make such a proposal, however, is evidence that the large-scale production of synthetic CVD diamond is more than just a pipe dream.

Three years earlier, *Science* magazine—one of the world's most visible and prestigious publishers of original scientific research—rated synthetic diamond as the "Molecule of the Year." "Diamonds may soon be everyone's best friend," wrote *Science*'s editor Daniel E. Koshland. "According to enthusiasts, synthetic diamonds have already or will soon appear on watch crystals, eyeglasses, optical instruments, audio speakers, fuel injection nozzles, turbine blades, scalpels, and semiconductor wafers, to name only a few applications."

Some of the first diamondlike carbon products indeed made it to the market by 1990 when Sony began marketing high-end speakers whose tweeters contained sound-generating diaphragms coated with diamondlike material. The diamond presumably

FIGURE 24

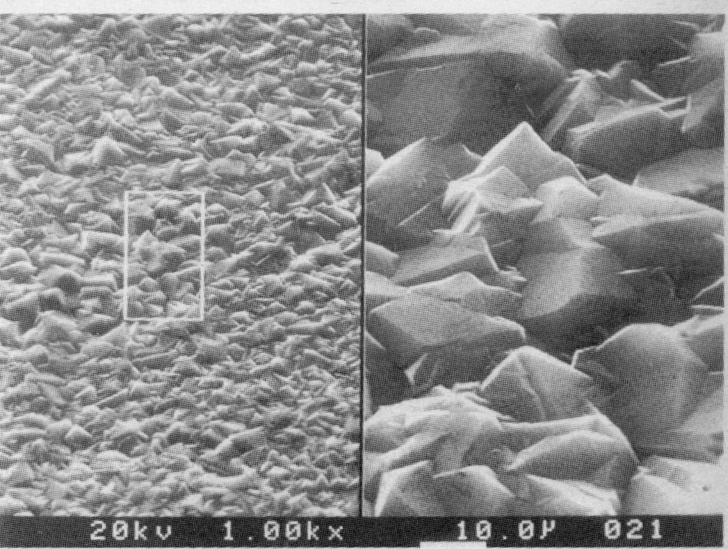

enables these components to improve sound quality by displacing audible distortions to frequencies beyond the range of human hearing. Seiko sells watches with scratch-proof diamond-coated faces. Crystallume, in Menlo Park, California, has marketed diamond-coated windows for infrared scanning systems, which are important in analytical instruments and missile-guidance systems like those used in the Patriot missiles. IBM scientists have looked to a CVD process to fashion light-filtering masks made of patterned diamond films, which they hope to use as a stencil for making smaller electronic circuitry. In turn, these speedier components would enable electronic engineers to build faster and more powerful computers and communications systems. The Norton Company, GE, and others have marketed heat sinks for hard-to-reach or particularly critical electronic components.

Fired by the promise of CVD diamond films, researchers and inventors are suffering from a new brand of technofever. "I'm addicted to diamond," admits DeVries, the retired GE diamond-research veteran. "There are a lot of us who have this addiction called diamond fever." The relative ease of getting into the fray has something to do with it. "You can almost make this at home with a microwave oven" says DeVries.[13] Several groups in Japan, the United States, and elsewhere have even used oxygen-acetylene torches, a $250 tool used for welding metal, to rapidly deposit diamond onto metal surfaces. Each acetylene molecule contains a pair of carbon atoms, the raw bits of diamond.

The relatively mild and easy-to-maintain conditions of CVD appear to be bringing diamond film into the club of readily available materials like glass, polypropylene, and aluminum. Yet even as synthetic CVD diamond makes its own bid to become a twenty-first century wonder material, many researchers are on the trail of materials that might contest diamond's preeminence. Some of these materials may already be here, germinating behind a veil of secrecy in government or industrial laboratories.

Then again, the lack of generally accepted claims of harder-than-diamond materials might mean that no one has pulled that off yet. After all, harder-than-diamond research is an especially tough business because no one thoroughly understands what makes materials hard. "Hardness is one of the oldest and yet most poorly understood of all of the physical properties of solids," according to the 1990 NRC report on the synthetic diamond technology.[14]

A variety of empirical tests provide simple hardness ranks for materials: what can scratch what or how much force it takes for a rod to dent a material. But these tests say nothing about how the underlying atomic and molecular structures of those materials relate to hardness. Hardness does correlate with such properties as the number of atoms or chemical bonds within a specified volume of the material: the more atoms or bonds the harder the material. But there are so many exceptions to these rules that they fail to explain hardness in general. Besides, correlations often turn out to have no universal significance.

Nonetheless, some theorists, including Marvin Cohen at the University of California at Berkeley, have examined hypothetical crystal structures made with carbon and nitrogen that might be harder than diamond. A hold-in-your-hand sample of such a material is eagerly awaited.

Another contemporary trajectory beyond diamond began in 1990, when GE rented out New York City's august 21 Club to announce that its researchers had developed a synthetic diamond that exceeds natural diamond's already world-record ability to dissipate heat by 50 percent. Moreover, the synthetic gems, which can take up to one week of press time to make, could withstand ten times as much laser energy as can natural gems.

Rather than using normal carbon sources, which contain mostly two kinds of carbon atoms called isotopes—carbon 12 and carbon 13—in a ratio of roughly one-to-ten, the GE team started with a carbon source enriched with carbon 12. The isotopic ratio was more like a hundred to one. The carbon isotopes have identical chemical behavior and bond into graphite or diamond lattices in exactly the same way. The only difference between them is that carbon 13 contains one more neutron in its nucleus than carbon 12, so it has a slightly higher mass. It turns out that a diamond crystal lattice made from carbon 12–enriched ingredients can conduct heat faster and withstand more laser energy than expected.

Decades earlier theorists had hinted at the possible benefits of these so-called isotopically enriched diamonds. One of them, Russell Seitz of Harvard University, claims to be the uncredited inspiration for the GE superdiamond. In 1990 he even bolstered his side of a very public priority argument with GE by waving Tom Clancy's book, *The Cardinal of the Kremlin*, at GE and the press. A technical cornerstone of the book depends upon an isotopically enriched diamond window for an X-ray laser. In the book Clancy actually acknowledges "Russ," a brilliant and quixotic college dropout in Cambridge, Massachusetts.[15]

Nobody had tried making the isotopically enriched diamond before the GE team because until recently most researchers expected that any enhancements in properties would show up only at exotically low temperatures. But the GE material struts its superdiamond stuff at everyday temperatures. These enhanced properties of isotopically purified diamond could be pivotal for the optical components—mirrors and beam-steering elements—of especially powerful free-electron lasers.

The price tag on GE's superdiamonds will scare away all but the most exotic and affluent customers—perhaps only military clients in search of components for new, more powerful lasers. But it also could prove a viable choice as a heat-management safeguard for delicate electronic devices in nearly impossible-to-reach places such as outer space or on the bottom of the ocean. Keeping orbiting or space-bound electronic chips cool with superdiamonds might eliminate the need for bulky cooling structures, which sometimes constitute half the weight of spacecraft. And every pound not shot into space can translate into thousands of dollars saved. For ocean-bottom communications lines, which often require relays and boosters, using superdiamond to rapidly shunt away device-degrading heat would minimize the need for costly and difficult maintenance expeditions.

The synthetic diamond tribe sees many facets in a gleaming future in which one of history's most prized and precious materials becomes an ordinary part of the technoscape.

BIOMIMETIC MATERIALS

The superlative properties of diamond were bound to attract the envy of materials makers who wanted to duplicate or surpass nature's own ingenuity. All of that from what might be the simplest of all materials. After all, diamond is made of one type of atom, carbon, each one bonded to four others into a completely uniform crystal. Other inorganic gems, including garnet, ruby, sapphire, and emerald—all of which involve more than one kind of atom to form more complicated crystals—have served as benchmarks in the materials engineers' unabashed quest to improve upon nature.

So far nature has a very comfortable lead over humans in engineering ingenuity. The earth's natural materials—among them spider silk, skin, wood, cotton, rubber, cockroach shell, tendon, bone, abalone shell, and ivory—are as miraculous in their capabilities as life itself. They are materials on which great and small civilizations have risen and fallen long before anyone had an inkling of how life makes these things.

Until recently, nature was the sole supplier, bestowing precious materials from its factory, the earth and its organisms. Acknowledging nature's genius, a cadre of materials researchers has been forging a newly recognized field, biomimetic materials research. Its practitioners aim to learn techniques from nature and then beat it at its own game, the way a novice humbly learns at the master's feet with an eye to superseding him later on.

The biomimetic credo is to learn as much as possible about the material structures within organisms and to exploit that natural technology into a scientific or industrial context. You might not want to make an artificial abalone shell, but you might want to mimic the abalone's method if making supertough ceramic products for armor or for lining the pistons in a car's engine.

The name "biomimetic" may be new, but the practice is not. Just a few years after Leo Baekeland's 1909 introduction of the first fully synthetic plastic, researchers in German chemical firms began turning out the first synthetic rubber materials, though falling prices of natural rubber rendered the man-made material economically inferior.[16] Until World War II the level of research on synthetic rubber followed the fluctuating prices of natural rubber. Natural rubber production always far outstripped the manufacture of synthetic versions, including neoprene, the rubber material that Wallace Carothers invented at DuPont.

That began to change dramatically in October 1942, when the U.S. government launched a synthetic rubber program (described by some as a forgotten Manhattan Project), one of this century's most impressive and unacknowledged successes of chemical engineering.

The program's charge was to secure a domestic source of rubber material for the war effort in lieu of the natural rubber supplies from Malaysia, which had been cut off by a strictly enforced Axis embargo. At the time there simply was no known substitute that could economically duplicate the combination of toughness, strength, elasticity, and durability of vulcanized natural rubber. Without rubber, however, the war's rolling machinery, with its belts that conveyed goods or translated power from a motor to a lathe or saw, and the tires of military vehicles would all come to an ominous halt.

This was a case where the fate of hundreds of millions of people depended on the speed with which scientists and engineers could learn how to mimic nature. Chemists had long known that the basic unit of natural rubber was a molecule called isoprene, units of which link into chainlike rubber molecules (polyisoprene), which in turn could associate with one another to constitute rubber.

In a matter of three years from the start of the synthetic rubber program, scientists and engineers working on the project had developed a superior synthetic rubber formulation—polystyrene/polyisobutylene—and had throttled production up from laboratory scales to pilot scales of several tons per week to nearly 1 million tons per year by 1945. This was a massive collaborative effort involving all the major rubber companies in the United States and enormous sums of money from the U.S. Government. Since then the synthetic rubber industry has grown into a vast enterprise producing well over 9 million metric tons of synthetic rubber, more than twice the amount of natural rubber. Synthetic rubber might be thought of as the most visible achievement of biomimetic materials even though it began well before the field even had a name.

It is easy to see the attraction of the field of biomimetics. Biology, after all, has accumulated several billion years' worth of awe-inspiring molecules and materials. A case in point is human skin, which remains a perfect fit as a person grows from infanthood to adulthood. During hard work, glands squirt sweat onto the skin's surface, where it evaporates, carrying away excess body heat. Nerve cells in the skin detect pressure, temperature, and the texture of objects. They also send pain signals to the central nervous system, which then activates muscles to get the skin away from whatever is causing the pain. Even when skin gets damaged despite these safety mechanisms, it repairs itself. All this from a thin, pliant, sensory covering that looks to the eye more or less like a homogeneous sheet. Bone, tendon, eye lens, horn, teeth, vascular tissue, beetle cuticle, cocoon silk, and many other biological materials have similarly awe-inspiring characteristics.

To Dan Urry of the University of Alabama, any material that can undergo three billion cycles of expansion and contraction without failing is a material worth emulating. That is precisely what vascular tissue does during a lifetime. Mehmet Sarikaya of the University of Washington is similarly in awe of the abalone's virtuoso ability to start with the stuff of chalk and antacid tablets (calcium carbonate), an intrinsically crumbly, weak ceramic compound, and create a beautiful iridescent shell tough enough to withstand hammer blows. Moreover, since an abalone grows its shell, perhaps it can teach engineers how to grow tough ceramics into complicated shapes like turbine blades. (See Figures 25–28.)

Julian Vincent, a British Zoologist at the University of Reading who brings a unique clarity and comedic spirit to the field, expresses the spirit of biomimetics this way: "Good scientists borrow ideas, but great scientists steal them." A spirited pedagogue, Vincent has been known to expose his slightly potted belly in airports to show how the sound of his alternately flexed or relaxed belly supplies rough information about the microstructure of fat and muscle tissue underneath his skin.[17]

His playful approach is not unique among his biomimetic colleagues. There seems to be a rare research spirit that the field brings out in its practitioners, a blend of reverence, delight, faith, and anticipation that their research could well lead to generations of high-tech materials with lifelike sophistication.

But all is not sweetness and light. In the early 1990s the ubiquitous Rustum Roy waged a rabid one-man war against the biomimetic research community, charging it with irresponsible self-promotion designed to wrest a larger slice of the research-dollar pie than it might deserve. Roy's aspersions led to published polemics between him and members of the biomimetics community in such journals as *Nature*, *Science*, and *Advanced Materials*.[18]

But controversy has not dimmed the ardor of the biomimetic warriors, who are still battling toward the holy grail of a new generation of biologically inspired materials that will rival biology's models in their versatility and practicability. Still according to Vincent's fellow zoologist Steven Vogel of Duke University, they should not necessarily seek to ape nature but rather to learn its design principles and apply them in new ways. Vogel points to the lesson of Otto Lilienthal, author of a nineteenth-century book *Bird Flight as the Basis of Aviation.* In 1886 he set out to achieve flight by emulating birds without realizing that they overcome the inherent aeronautical instability of their bodies with fantastic neural control. "Lilienthal designed very birdlike hang gliders and was killed in one of his thoroughly unstable craft," Vogel pointed out in his own book *Life's Devices*.[19]

There are other caveats for eager buyers of biology's evolutionary brand of materials engineering. For one, biology has done its engineering within small ranges of physical conditions that can continually support life. "This is equivalent to transforming a motorcycle into a car without ever losing the benefit of transportation," Vogel writes. Moreover, the very logic of evolution pushes not toward optimization but toward perpetuating evolutionary traits that are merely adequate for the survival and reproduction of a population under prevailing environmental conditions. So working with available raw materials, the process has led to hard, soft, stretchy, sensitive, and other kinds of materials that work just fine. They are not necessarily the best materials that could have been made from those ingredients, but they obviously are sufficient.

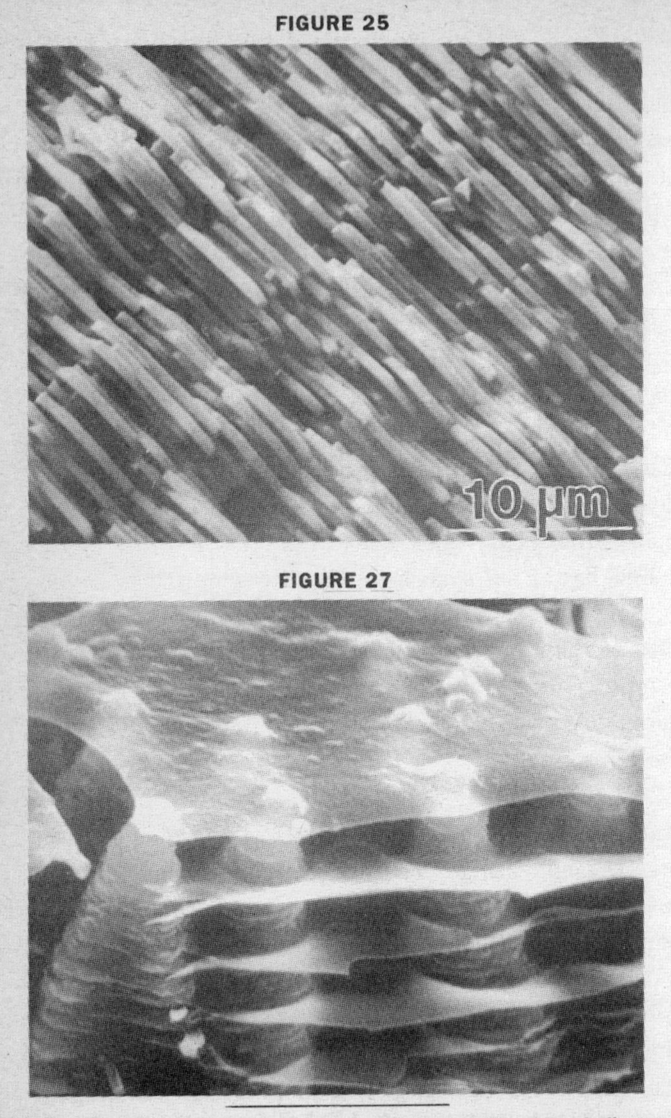

Molluscan Materials Engineering. In the tough inner layer of its shell, an abalone lays down calcium carbonate crystals in a bricklike fashion with an organic mortar composed of carbohydrate polymers and other biochemicals (Fig. 25). The resulting material is both strong and fracture-resistant, since a crack has to take a torturous course through the layered structure instead of speeding through in a straight line (Fig. 26). This is an image of the leading edge of a shell growth of an abalone (Fig. 27). The conelike pillars

FIGURE 26

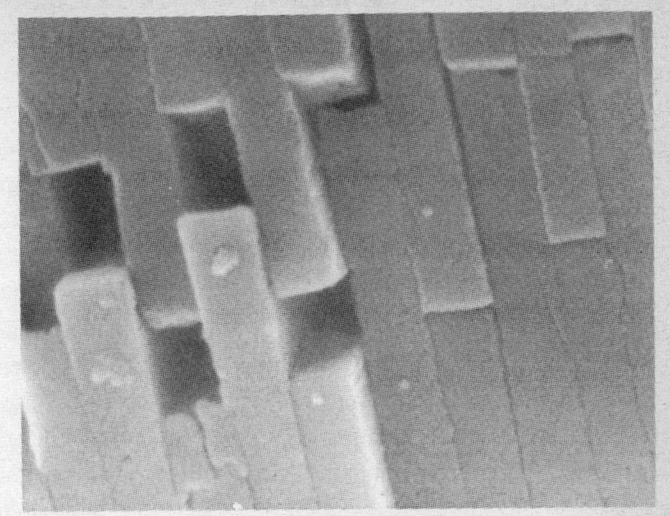

FIGURE 28

grow and merge together to yield a full shell. A synthetic analogue of the brick-and-mortar shell structure consists of multiple layers made of a boron-carbide-polypropylene mixture alternating with thinner layers of polypropylene (Fig. 28). Compared to a monolithic piece of boron-carbide, the mutlilayer architecture yields a material that is considerably more fracture-resistant.

MEHMET SARIKAYA

The scientist's freedom from nature's constraints is what inspires biomimetic researchers the most. After all, human engineers can opt to use cryogenic or kiln temperatures, extremely low or high pressures, and highly acidic or basic ingredients. It follows therefore that they can try to develop, say, faster means of growing biologically inspired ceramics instead of relying on the slow growth common in life's own methods for manufacturing bone and shell.

This potential has not been lost on military research managers. "Nature has these wonderful solutions and exquisite structures that go far beyond anything we have now," Michael Marron, then a program manager in the Office of Naval Research, told *Science* magazine in 1991. Biological materials could well serve up the concepts leading to more durable, survivable materials and structures for naval use: for example, new underwater adhesives (perhaps based on the "byssus" that mussels use to attach themselves to rock), anticorrosion coatings, or lightweight yet steel-strong composites that make the shells and exoskeletons of certain beetles so strong.

Nor has the potential been lost on the civilian business sector. Chemical giants like ICI, Hoechst Celanese, starts-ups like Protein Polymer Design in San Diego and Adheron are all banking on biology to point ways to new markets and profits. Their targets range from bioceramic implant materials for grievous bone injuries, composites for aircraft, and new fibers inspired by spider silk.

The richness of biological materials comes from their hierarchical design, a point that Eric Baer (a colleague of CVD diamond expert John Angus at Case Western University) has been making for years. Living things exercise exquisite control over chemical composition and structure at every structural level, from atomic and molecular components to intermediate structures such as fibers and crystal grains to visible tissues and components such as tendons, eye lenses, arteries, skin, and bone.

Baer finds collagen, a protein, to be an instructive example since biology puts it to multiple uses in connective tissue, tendon, tooth, and bone.[20] Collagen is a chainlike protein molecule made up of linked molecular building blocks called amino acids, which in turn are made of about twenty atoms of hydrogen, carbon, oxygen, nitrogen, and sometimes sulfur. Each protein molecule serves as a molecular strand. These strands twist into stronger bunches of many strands. These bunches twist into even stronger fibers that together twist into strong and tough collagen cables. In bunches these cables form anatomical parts like tendon. In bone the collagen cables strengthen the hydroxyapatite mineral that makes up the bone's bulk. Combined with biological minerals such as the hydroxyapatite in tooth dentin, they toughen teeth. That is why Paul Calvert of the University of Arizona has been using rat teeth as a model for developing synthetic composites using synthetic polymers instead of collagen and titania or other tough ceramics for the bone mineral.[21] (See Figure 29.)

Calvert admires nuts, too. This is what a macadamia nut means to Calvert: "The shell of a macadamia nut is an 'isotropic wood' structure, comprising bundles of cellulose fibers arranged to be locally parallel but random over longer distances. Macadamia shells are notoriously tough, and this can be attributed to the random, fiber-reinforced structure." Presumably materials scientists would be able to increase the toughness of fiber-reinforced polymer composite materials by duplicating the macadamia nut's various levels of microstructure in the bulk of the synthetic counterpart. No one has figured out how to do that yet, Calvert has noted.

Mehmet Sarikaya found his teachers off the coast of Baja, California: red abalone shells, which are nearly as fracture-resistant as synthetic ceramics such as zirconia, carbon boride, silicon carbide, and silicon nitride.[22] And yet the shell is made almost entirely of calcium carbonate, the same stuff as chalk and most antacid pills, which crumble at the flick of a fingernail.

The enhanced properties of the shell derive from the way the mollusk's cells build calcium carbonate into crystalline microbricks with chemical components derived from the seawater around them and cobbled into a microscopic brick-and-mortar structure. The shell consists of two main layers in which calcium carbonate is arranged in two different kinds of crystal structures. The outer layer consists of calcite; the inner, tougher, iridescent nacreous layer consists of aragonite. The two layers differ only in the relative spacing and angles of the crystal's calcium and carbonate ions. In the inner layer the calcium carbonate forms layers of microbricks within which each single crystal is between a quarter and half micron high, or about one one-hundred thousandth the thickness of a building brick. Between each of these layers—and between bricks on the same layer—is a tenfold thinner layer of "organic mortar" composed of carbohydrates and protein molecules whose precise composition remains unknown.

The structure means that cracks get stopped in their tracks as they attempt to take cumbersome, tortuous routes around the single crystal bricks. Under high magnification, the researchers could see stringy ligaments, derived from the organic mortar, bridging adjacent bricks that had separated in cracks. The ligaments, Sarikaya says, resist and diffuse crack energy. In 1994 Greg Olson, a materials researcher at Northwestern University, adopted these observations as a lead for making steels that he and his students hope will heal themselves when too much strain causes their microanatomy to form cracks.[23]

The Hierarchy of Living Stuff. This classic diagram of the biomimetics field illustrates how tough, elastic tendons emerge from a hierarchy of structural levels beginning with tropo-collagen molecules that twin into microfibrils, which pack into subfribils, which in turn pack into fibrils, which form into fascicles, which combine into a tendon that you can see. Similar atom-or-molecule-to-total-thing sequences underlie nearly all materials.

JOHN KASTELIC AND ERIC BAER

In 1989 Ilhan Aksay, Sarikaya's sometime coworker and colleague, had taken this remarkable structure as a model for making hybrid materials known as "cermets," which are made of one ceramic component and one metal component. But instead of mixing the components without much concern about the microstructure, the researchers thought of their boron carbide as the calcite bricks in the abalone shell and the aluminum as the organic mortar. Compared to B_4C/Al cermets, in which there is no attempt to mimic biological microstructures, theirs was 30 percent more resistant to fracture, and they could see from microscopic examination of deliberately created cracks that abalonelike toughening mechanisms appeared to be at work. This enhancement occurred even though the layers of the cermet material were tens of times thicker than the calcite layers in the abalone shell. This told the researchers that there was plenty more improvement in store for the clever materials researchers who could more closely replicate the shell's structure.[24]

The lesson is to create laminated structures with highly

FIGURE 29

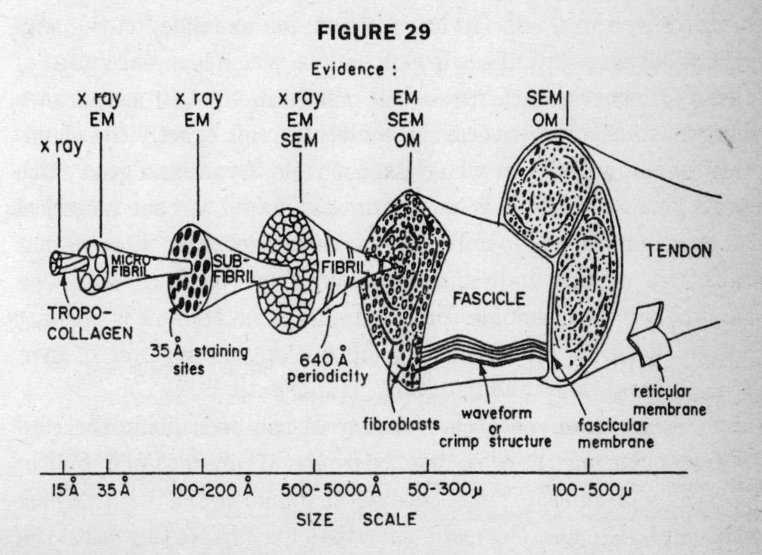

ordered microarchitectures. The principle is to use one material that is hard and one that is softer yet capable of forming strong adhesive interfaces. Using that molluscan design principle with materials that are intrinsically much harder than calcite, such as boron nitride or titanium carbide, remains a possible route to the toughest and strongest ceramic materials ever made.

Even if researchers flesh out such possibilities in a laboratory, a small-scale success may not translate to the scaled-up production required to make a new generation of armor, manufacturing tools, or engine parts. This kind of commercial anxiety attends the discovery of nearly all advanced materials. They may look good on paper or even in the lab, but that is not enough for companies to risk the vast sums of money, credibility, and marketing energy that it takes to convince potential customers to switch over to a product inspired by a mollusk shell.[25] A material can have world-record performance, but if you can't make it cheaply enough, then it will remain a laboratory curiosity.

Others in the biomimetics tribe have been exploring biomineralization. Chemist Stephen Mann of the University of Bath is one of them. He has focused, for example, on the way the calcite crystals of sea urchin spines, which look like spires of a Gaudi church, nucleate on cell membranes.[26] Cell membranes are made of oily hydrocarbon molecules, not exactly the chemical venue where you would expect rigid crystals to grow with any kind of finesse. What Mann has found is that electrical charges on the cell membranes serve as nucleation sites so that a bunch of tiny individual crystals grow perfectly into one another to form a single complicated crystal. He now is pushing ahead with his research, focusing on various systems of synthetic membranes and inorganic components.

A group of researchers in Israel has been doing the converse—learning how to use coatings of predesigned hydrocarbon molecules on small crystals to influence the way in which the crystals grow. Normally scientists are unable to control the way in which molecules aggregate into specific crystal shapes such as flakes or cubes or needles. Yet the ability to control relative growth of the different faces of a crystal could have a lot to do with controlling ice formation on airplane wings, reducing damage to tissues preserved by freezing them, or even growing high-tech crystals for detectors or electronics. What it takes is knowing the precise physical, chemical, and electronic contours of the crystal's various faces and the ability to synthesize chemical compounds that nestle into specific contours.[27]

One of the more daring arenas of the biomimetic community centers on the softer world of protein-based polymers, the stuff of blood vessels, hair, silk, and muscle. One of the special attractions here is that the ingredients of these materials are encoded in genes, which molecular biologists have spent the last forty years learning how to manipulate. Not only might biological polymers serve as design models for new high-tech polymers, but the machinery of life itself might also be commandeered to carry out the redesigning and manufacturing.

As for models of new polymer materials, Dan Urry of the University of Alabama has been pushing the envelope as much as anyone. His successes were notable enough to earn him *R&D* magazine's Scientist of the Year honors in 1988.[28] Urry, perhaps echoing some of Roy's caution, thinks that the living kingdom takes too long to construct the hard parts of its anatomy.[29] It takes a sequoia many hundreds of years to create the several thousand tons of its bulk. That does not synchronize well with the needs of modern materials industries, which might have to deal in tons of materials per day.

He remembers the time in 1978 when he discovered the repetitive structural motif—a sequence of five amino acids found in elastin, an elastic protein-based material—that seems to at least partially account for the wondrous qualities of blood vessels, lungs, and other tissues that repeatedly stretch and relax.

He has since found ways of chemically modifying these

molecular constructions, which he then can cast into sheets or draw into fibers. One of his sponsors, the Navy, has been interested in his work since it could lead to resorbable surgical implants that might prevent the painful and often life-threatening adhesions that occur on and near surgical wounds after operations. In this application it is the protein-polymer's compliance and resilience that are important, and so it serves its function passively but efficiently.

These materials continue to take Urry's breath away because of how they act. He has learned to modify them so that they respond to various environmental cues such as changes in acidity, temperature, voltage, salt concentration, and pressure. When exposed to a flash of light, a foot-long string of one version of this polymer holding up a weight will contract and raise that weight by several inches in what appears to be an almost lifelike response. Urry can show off similar effects with different versions of the polymer immersed in fluids or in spaces in which he can control the temperature: here, potentially, are polymer materials working as synthetic muscles, either for bionics (human applications) or robotics. Here, at the crossroads of the laboratory and the functioning of the human body, we find contemporary biomimetic materials research fulfilling the highest hopes—and, perhaps, some of the deepest anxieties—of a culture rapidly narrowing the gap between the organic and the synthetic, between humanity as the product of natural design and nature as the product of human design.

SMART MATERIALS

Urry's active polymers are a many-talented lot, chafing at any hint of biomimetic typecasting; these versatile materials are equally adept as performers in another sexy and contemporary genre known as "smart materials and structures"; metals, ceramics, and other components, including microprocessors, are also members of this astute and polymathic league of inanimate structures.

The field began quietly in 1959, about the same time that Advanced Research Projects Agency (ARPA) was drawing up legal papers to create the first three Interdisciplinary Laboratories in an attempt to institutionalize the modern form of materials science and engineering. That was when William H. Armistead, an executive at Corning Glass Works, challenged S. Donald Stookey, the company's most innovative researcher, to create something truly remarkable. Armistead, who later wrote a brief foreword for Stookey's quirky and fascinating autobiography, wanted a new kind of glass that would automatically darken in the light and lighten again in the dark. Somehow light would have to trigger light-sensitive molecular changes.

Stookey's solution came partly from the light-triggered chemistry of photography that had made Eastman Kodak so famous. A decade earlier, Stookey and some Corning colleagues had experimented with additives that caused a change in the internal structure of glass in response to light exposure. Stookey recalls in his autobiography that the government delayed the patent for one of these for a few years because of its potential use in carrying secret messages. "Exposure to sunlight or ultraviolet light through a stencil or a photographic negative imprints an invisible latent image which can at any later time be developed to a visible image when the glass is heated to 600°C," Stookey explains.[30] Two spooks from the Office of Strategic Services showed up to collect some samples, though Stookey never learned what they had done with their booty.

For the photochromic glass that Armistead challenged him to make, Stookey added silver halide spiked with copper halide to the glass formula. These salty inclusions dispersed through the glass like tiny crystalline islands in a less ordered molecular sea, like grains of salt in pudding. The crystallites were so small and transparent that they hardly absorbed or scattered light.

The material brilliance of this glass formula is a tango of light and chemistry. Energy from light knocks an electron from the copper ions present in the tiny salt islands, and nearby silver ions pick up the liberated electrons to become neutral silver atoms. These neutralized silver atoms then form millions of tiny light-blocking specks throughout the glass, just as they do in photography to form an image. When the light dims, the silver atoms in the clumps return their borrowed electrons to the copper ions, and the clumped silver atoms disband again, which means the glass clears again.

By the 1970s photochromic glasses had made their way into commercial items, mostly eyeglasses with a self-adjusting, variable transparency. According to Donald S. Trotter, a member of the contemporary generation of Corning's glass and ceramics researchers, if researchers can tame their temperature sensitivity, which makes these smart materials darken more intensely on colder days, they could find larger-scale applications in such places as buildings and automobiles.[31]

At the end of the 1970s, even as photochromic glass was flanking the noses of thousands of eyeglass wearers, researchers at NASA began talking up ideas that eventually developed directly into the field of smart materials.

Richard Claus, a fiber optics expert at the Virginia Polytechnic Institute and State University in Blacksburg, recalls the time this way: "The basic concept of smart materials evolved back in the late 1970s. I was working at NASA at the time, and we were trying to develop 'nerves' and 'nervous systems' for materials."[32] Spacecraft were expensive conglomerations of the highest technology, sometimes taking human beings across the atmospheric threshold into space. The researchers wanted better ways of testing materials and components of spacecraft without always having to take a good portion of the spacecraft apart, which itself could have ill effects on the materials involved and add risk to various missions. "The idea we had was to develop nondestructive sensing applications [essentially, means of looking inside structures without touching them] to tell when the structure would not be able to perform the way you would like," Claus said.

Like the nerves in human skin that send easily interpreted signals of pain in response to a wound, the sensors that Claus and his fellow NASA researchers were talking about would detect cracks, delaminations, unusual porosity, excessive strains, and other flaws that could easily lead to front-page catastrophe. One possible way of creating these artificial nervous systems was to embed optical fibers into the spacecraft's skin material. Changes or blockages in the light passing through the fiber would be the smoking gun indicating that something in the skin was no longer right.

No one had yet been using terms like "smart" in reference to spacecraft rigged with sensors. Claus dates the beginnings of that kind of talk to the early 1980s, even tentatively tracing the origin to a closed-door meeting of Air Force planners known as Forecast II, the sequel to Forecast I, which was convened soon after the emergence of jet technology during World War II. World War II had rendered air power of strategic military importance, and from then on, any technological edge in aerospace technology could change the balance of power. That was the idea behind Forecast I and Forecast II. The charge of these meetings was to outline where the Air Force was going in the next twenty years in terms of materials, structures, and technologies. "They came up with the top ten technology topics," Claus said, and one of them was "smart materials and structures."

The idea here was to have a plane with a "smart skin" that could sense the aeronautical conditions—things like temperatures, pressure, wind speed—as well as internal strains and maybe even the onset of damage. Maybe, like living tissue, the smart skin could even automatically compensate for these changes or even for a direct antiaircraft hit through a wing. That could give a pilot that extra second or the freedom to concentrate on more important things, the sort of advantages that can mean the difference between life and death for the pilot and success or failure for a mission.

Initial work focused modestly on smart skins in which optical fibers would serve as damage sensors. "Gradually people warmed up to the idea," Claus said. The main developments revolved around using optical fibers as sensors embedded in composite materials akin to fiberglass but far stronger.

As more researchers with different backgrounds began sniffing around in the field in the 1980s, the concept of smartening up materials and structures began expanding from its roots in "smart skins" for aircraft and spacecraft. Besides glass fibers as sensors, any material capable of sensing something became part of the mind-set. The idea was to develop a multisensory capability modeled after the human body's ability to monitor what is happening around it, on it, and within it. This is how a human body can react to a lion coming at it, dangerously high carbon dioxide levels in blood, and the scrape wound on a knee. Any material with some measurable property that changes in response to stimuli in a mathematically predictable way became part of the endeavor. Prototype hardware was validating the concepts while revealing technical problems, such as the way optical fibers can weaken composite structures, making them more susceptible to the damage they are there to detect.

Technophiles have a knack for quickly becoming jaded. So what if you can rig a structural beam or aircraft skin with sensors—what else can you do? Enter Craig Rogers, who was with Richard Claus at the Virginia Polytechnic Institute, where the country's first research center devoted to smart materials and structures opened for business in 1987. And that's about when

---

Smart Stuff. By building the structural members of vehicles, airplanes, and bridges with materials that can sense environmental conditions as well as materials that can respond to those conditions, engineers have designed materials systems such as concrete and airplane skins that can behave "intelligently" by alerting humans when dangerous situations arise or by changing their properties in accordance with changing circumstances.
CRAIG ROGERS

the smart materials and structures concept began taking off.[33] (See Figure 30.)

Pivotal to this expansion were piezoelectric ceramic and plastic materials, which respond to mechanical deformations from pressure and vibration by creating a change in the voltage across their bulk. Piezoelectric materials work in reverse, too, a feature that sonar developers exploited during World War II to develop submarine detection systems of unprecedented aural sensitivity. Just as they respond to mechanical deformation by creating an electrical signal, an electrical signal fed into them makes them deform or vibrate. And that means that they not only can sense things, but they can also do things. Like muscles, they can make things happen.

These people stopped looking at walls, girders, beams, airplane skins, hulls, windows, and other structures as essentially passive things whose only function was not to fall apart. "Let's put some sensors and actuators into them. Let's put a control loop around it, and let's do things like damp out vibrations in structures, let's eliminate wing flutter, let's make the material

FIGURE 30

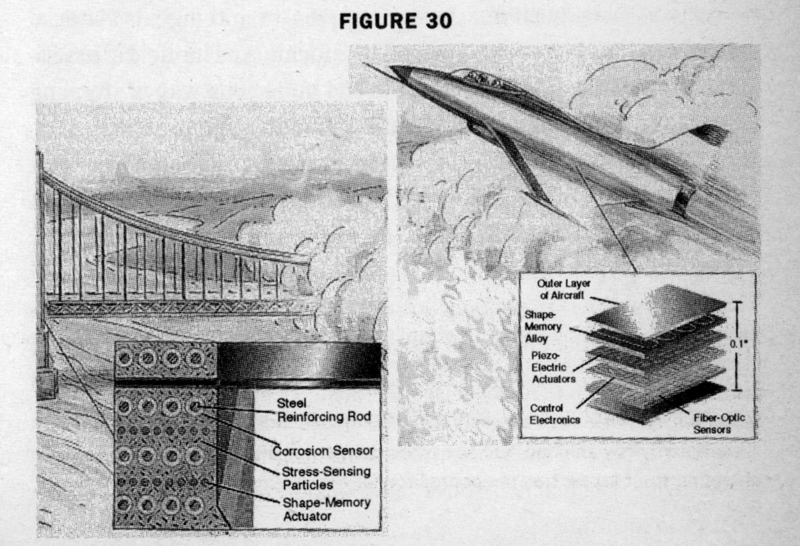

respond to the environment"—these became the catch phrases of "smart materials" researchers.

By the 1980s the diverse lines of smart-materials research had begun to converge enough to warrant the honorific "discipline." In 1988 a few hundred of these scientists and engineers convened the first meeting devoted to this area. It kindled a community sensibility that has been growing ever since. After that there was a meeting devoted to the field every half year or so, and other engineering and optical conferences began to have special sections. In January 1990 the premier issue of the field's first journal—*Journal of Intelligent Materials, Systems and Structures*—rolled off of the presses. Craig Rogers, its editor, became one of the field's most visible champions, even appearing in a 1991 PBS special called "Miracles by Design." Within a year *Smart Materials*, another new journal devoted to the field, this one edited by Claus, began publishing.

Rogers, who now is dean of the college of engineering at the University of South Carolina, takes the biological analogy very seriously. At more than one scientific conference, he has flashed a slide showing his young daughter next to a "smart cylinder" that actively resists buckling: two examples of intelligent material systems. Biology is just part of the formula. The field's practitioners have developed an eclectic and audacious way of thinking that draws upon any useful science or engineering practice.

Like their biomimetic brethren, smart materials researchers look at biology for metaphors. For the prototypical intelligent material system Rogers envisions structures that have sensory systems that feed input about the internal state of the system and its external conditions into some kind of computational component. This component—probably a microprocessor or network of microprocessors—would interpret these signals to plan out and execute a response by controlling an assortment of musclelike actuators. Neural network processors might actually enable the structures to learn proper responses and even adapt to changing situations.

With that basic schema smart materials researchers have embarked on a variety of projects: rope that changes color according to how much strain its polymeric fibers are enduring; adaptive camouflage materials that sense the surrounding landscape and then alter the camouflage's pattern of colors and shades like the skin of a chameleon; and smart beams, hulls, rivets, and engine mounts, to name only a few.[34] Smart beams, as Rogers sees them, are structural members that can change their stiffness depending on the load or the kinds of vibration they are subjected to. And that is a route to making things last longer. Aeronautical engineers aim to make wings that change their shape, their aeronautical contours, in accordance with the prevailing conditions; they thereby hope to steer the craft without the need of flaps. According to Julian Vincent of the University of Reading, whose work straddles biomimetic and smart materials, "We are trying to build intelligence into the constructed world."

When you ask Craig Rogers what good he thinks intelligent material systems might be, he squints his eyes as he scans over a mental scenario that he thinks about every day and believes will play out in the objective world beyond his own hopes. "The concept of instilling lifelike functions in inanimate objects and artifacts such as advanced composite materials and home appliances seems to be a vision more akin to science fiction than present-day realism," he acknowledges. But he stresses that an entire community of researchers has collectively come to this view, that it "is not the result of personal vision resulting from some manifestation of divine intervention." Yet he still won't shy away from using prophetic terminology. Without remorse he has committed the following to paper: "Intelligent material systems will be manifestations of the next materials and engineering revolution—the dawn of a new materials age."[35]

He expects this "revolution" to engender major changes in the philosophy of engineering design. By replacing material brawn with intelligence, engineers may no longer have to build structures with tons of extra concrete, steel, and other kinds of weighty materials to assure structural integrity. An intelligent crosswalk or dam or wall could parry unusually strong strains or vibrations, say during an earthquake, by sensing and responding to the challenges the way a standing train commuter bends and twists his knees and ankles to ride out a bumpy trip.

In time engineers will be able, in effect, to take a structure's temperature and pulse. "We will soon have the opportunity to ask structures during their life how they are feeling or where they are hurt," Rogers says. "If you embedded fibers in materials from the time they were manufactured until the time they were retired, you would be able to watch the material be fabricated, you would be able to look at it during its service lifetime, always monitoring for things like strain and temperature and damage."[36] This is the kind of capability that engineers and executives of aircraft manufacturers would love to put into their products.

Though the idea of smart materials and structures may have had its origins in Air Force discussions, other federal agencies have set their sights on smartening up their hardware. The Navy has long imagined submarines with hulls that could sense the sonar surveillance of enemy observers and quiver enough to either absorb the sonar energy without sending back an echo or send back a deceitful signal that would effectively cloak its whereabouts.[37] This would be yet one step up from the advanced silence of Tom Clancy's *Red October*.

Besides aerospace and submarine settings, the smart materials and structures idea is infiltrating other more mundane areas. One company has made smart rivets whose straight shafts bend back over the part to be secured by the application of heat rather than force. Another markets orthodontic wire made of so-called shape-memory metal that never has to be adjusted once it is programmed to coerce into proper alignment the teeth it is attached to. A high-end automotive use for these alloys is in switches that flip headlight assemblies in and out of the hood of sports cars.

The magical features of these alloys is that they can be programmed to assume a particular shape at a particular temperature even after having been deformed into a very different shape at another temperature. It is possible to make a coil of properly programmed metal that would unfurl to spell out your name when heated with a hair dryer. The underlying principle resides in the arrangement of the metal's crystal grains and the kinds of crystallographic shifts that occur within these at different temperatures.

OTHER SIGHTINGS

Synthetic diamond, biomimetic materials, and smart materials are just a few of the species emerging from the wilds of materials science labs. The dedicated observer, however, will catch sight of even more recent and exotic creatures of the scientific imagination.

Fullerenes, for example, are relative newcomers, cousins in carbon of the more established synthetic diamond. Their most famous representative is a ball of sixty carbon atoms that form into a soccer-ball-shaped molecules known as buckminsterfullerene (C_{60}) or buckyballs. Its carbon atoms literally bond into the hexagonal and pentagonal facets of a soccer ball. As far back as the 1960s, theorists had postulated their existence, or perhaps wished their existence because of their extreme display of molecular beauty. In the 1980s experimentalists with sophisticated analytic instruments began capturing wispy, blurry snapshots of their actual presence. Then, in the fall of 1990, a team of German and American researchers found a way to make noticeable amounts of them by sending an electric spark between two graphite electrodes.[38]

Since then fullerenes have been all the rage. Researchers have made a wide variety of fullerene balls and cages, including C_{70}, whose structure resembles a rugby ball, and other much larger ones involving hundreds, even thousands, of carbon atoms. There are nested fullerene structures with carbon balls inside of carbon balls creating structures resembling onions. A subpopulation of fullerene researchers focus on all-carbon buckytubes, which also often show up either in single shell or nested forms, this time as tubes within tubes within tubes. Experimentalists have verified that some of these buckytube structures can be electrically conductive, which means they could become molecular wires or even switches for electronic devices even more miniature than today's.

Richard E. Smalley of Rice University is one of the field's pioneers who won a Nobel Prize in chemistry in 1996 along with fullerene pioneers Robert F. Curl, also of Rice, and Sir Harold W. Kroto, of the University of Sussex in England. In 1985, he saw some of the first glimpses of C_{60} while vaporizing carbon inside vacuum chambers in the hope of learning how smaller clusters of atoms—a state of matter between atoms and big hunks of stuff—behave. Consider a flat fullerene sheet, also called a graphene sheet, which is what curves into balls or rolls up into tubes. "A graphene sheet is really a membrane, a fabric, one atom thick, made of the strongest materials we expect will ever be made out of anything, which is also impenetrable," Smalley told a gathering of chemical engineers in Houston in early 1996, "and now we realize it can be wrapped continuously into nearly any shape we can imagine in three dimensions. This has got to be good for something!"[39]

Smalley and others envision the buckyballs and buckytubes and other shapes hewn from graphene sheets as tinker toys for building nanoscopic constructions. If you hang metallic catalysts on these or enzymes, you might end up with a material for carrying out industrially or medically important reactions.

This brave new world of curved and tubular carbon could,

like the previously known forms of carbon, spawn a diversity of useful new materials. After all, diamond doubles as a mineral symbol of love (a public relations creation by DeBeers that has earned the company billions of dollars per year) and as a pivotal material in mass manufacturing. And graphite is the stuff of pencil "leads," dry lubrication, and reinforcement for composites.

Like biologists discovering a brand-new animal species, fullerene researchers have been prodding, modifying, and poking fullerenes and related molecules just to see what the molecules do. They have crystallized them along with ions of rubidium to create forms that are superconductive. Israeli researchers tried to hook a fluorine atom to every one of the sixty carbon atoms with the so far unrealized hope of coming up with a super teflon. The researchers got close to achieving the chemical goal, but they didn't end up with anything that seemed like teflon. Researchers at Northwestern found a way of using fullerenes as a kind of carbon adhesive that helps diamond films stick better to other materials. Another group of Northwestern researchers found a way of casting the fullerenes into films that could be quickly switched into a light absorptive state when hit with a laser beam, just the sort of thing needed for the nightmarish military scenario of battlefields thick with potentially blinding laser beams.

Still another tribe of researchers worships at the shrine of newer and better semiconductor devices; their religion of smaller and faster will, they hope, usher in an electronic New Jerusalem. Silicon has been and remains the stuff of the electronic infrastructure that innervates modern society so thoroughly now that its sudden disappearance would be the equivalent of a social nervous breakdown. Yet silicon, when properly tainted with charge-friendly impurities such as boron and phosphorous atoms, is only one member of the class of materials known as semiconductors.

You also have the "three-five" and the "two-six" semiconductors, which are made by mixing and matching the elements of the periodic table that lie within the table's corresponding columns. If you build crystals out of one part gallium (an element in column three of the periodic table) with one part arsenic (an element in column five), you get what many researchers hope will be the next-generation electronic material, gallium arsenide (GaAs), the preferred material for Cray supercomputers, since electrons move more swiftly through it than through silicon. Moreover, GaAs is host to optical as well electronic activity. When electrons and positively charged carriers of current called holes (because they are really the absence of electrons) meet inside the material, they can recombine with one another in a light-releasing embrace.

Each of these three-five or two-six materials—some made of three or four elements—has its own characteristic band gap, which refers to the energy difference between their electrons in an excited and mobile state and those same electrons in the relaxed or valence state, in which they remain hitched to a particular atom or group of atoms. What electrons and photons do inside of these materials depends on the size of the band gap. Some of these materials detect radiation of various wavelengths by converting incoming photons into discernible flows of electrons. Others turn electrical energy in the form of moving electrons into light, which is why these also are known as optoelectronic materials. It is this light-emitting property that has put gallium arsenide, not silicon, at the heart of laser scanners in CD players that bounce light off of the little hills and troughs on the plastic compact discs.

Despite that material wizardry, the running joke amongst its developers is that "gallium arsenide is the next-generation material: always has been, always will be." The pessimism comes from the difficulty and expense of producing it with the high quality necessary for making microelectronic circuitry. It took years of development and hard work to develop a reliable method of growing silicon crystal boules from which technicians could slice wafers that could be diced into millions of silicon chips and then patterned with microcircuitry to become the heart and soul of modern technology. But silicon crystals are made of one element, not two or more like gallium arsenide, or gallium aluminum arsenide, or indium phosphide, cadmium mercury telluride, or any other combination of elements in columns two, three, five, and six that physicists know will form a new electronic, optoelectronic, or photonic material. Still, there are devotees of each of these elemental combinations.

Other hot materials and trends du jour: "liquid crystal polymers" augur a generation of flat computer-screen displays or even television screens that roll up like posters; like gallium arsenide, electroluminescent polymers can be made to emit light when stimulated electrically—instead of electrons and holes meeting like lovers, the electricity riles some of the polymer electrons to excited states that vent by way of light emission; diagnostic-minded researchers are training the vast and powerful arsenal of analytical instruments on the material microstructure of the wood in a Stradivarius violin or in a bronze Chinese bell to reveal hard-won craft secrets that long predate the scientific acceptance of the reality of atoms; metallurgists are finding ways of combining metals in proportions and microarchitectures that will yield the right stuff for technologies like an aerospace plane that flies twenty times the speed of sound.

The beat goes on: some scientists are beguiled by the prospect of making ceramic objects like plates, spark plugs, and the tiles lining the Space Shuttle's exterior by using nanoscale particles of metal or ceramic particles that are comparable in size to viruses—the tinier the particle, the more surface it has relative to its interior, and most of a material's interesting features happen at its surface. Ceramics made with these tend to sinter or consolidate at temperatures hundreds of degrees

FIGURE 31

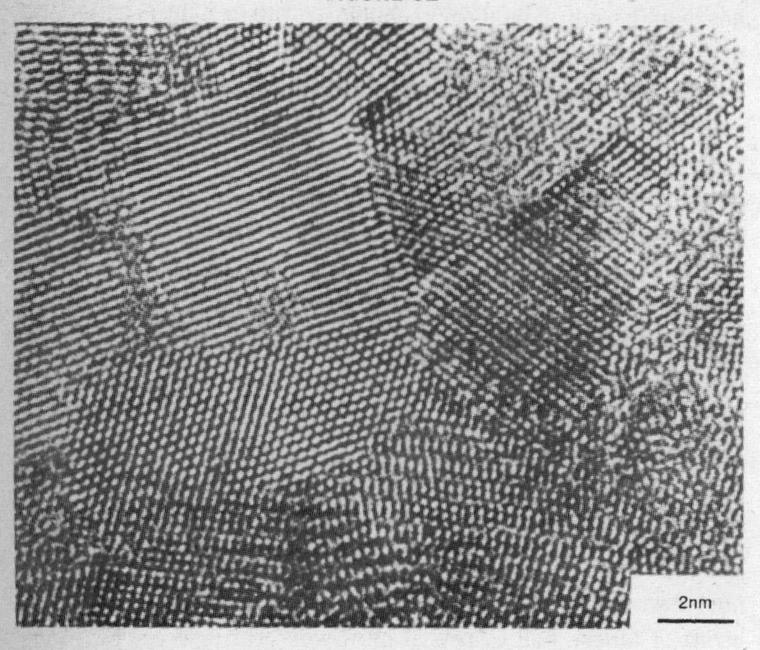

Very, Very Tiny Grains. Each of the discernible regions in this high-resolution image (from a transmission electron microscope) corresponds to a tiny grain of palladium metal. The grains are on the order of several nanometers (billionths of meters) in height, length, and/or width, which is why materials made with grains this size are referred to as nanophase materials. Researchers have shown that the much smaller grain sizes of these materials (compared to the grain sizes of powders used to make conventional metal or ceramic materials, whose grain sizes often are hundreds or more times larger) lead to important advantages. One of them is the ability to make ceramic components in complicated shapes without the difficult and expensive machining that so far has kept ceramics out of high-volume applications where their resistance to heat and wear could make a big difference.
ARGONNE NATIONAL LABORATORY

cooler than with standard powders. And they are more easily shaped. These are the kinds of features that can save millions of dollars in large manufacturing operations. (See Figure 31.)

There is a computer-wielding, acronym-happy tribe that lives in a world of theories and equations: quantum mechanical models that describe where and how electrons move and differential equations that account for the way a material's atoms and molecules move and jostle at different temperatures and pressures. They sometimes taunt their dirtier-handed brethren by predicting new material structures that will be in some way superior to any other material on earth: make a certain arrangement of silicon and nitrogen atoms and you well get stuff that is harder than diamond; make an odd ring of oxygen atoms and you will make stuff that can pack more propulsive power than any known fuel.

Some trends solidify; others evaporate as quickly as dew on a summer's morning. The field as a whole, however, seems on the verge of a critical mass of human and technical capability that will result in an explosion of technological bedazzlement. Rousing insensate matter into something like intelligent behavior is not simply science, but an art as well: consider the rich palette the researcher has in the periodic table, whose elemental ingredients he or she can combine into an infinite variety of molecular shapes and sizes and chemical blends. The resulting material portraits depict our future. Aping nature or outsmarting it, man-made matter matters as never before.

PART III

Materials by Design

Materials by trial and error: even an *Australopithecine* protohuman living 2.5 million years ago would have been justified in saying, "Been there, done that." But inventing new materials by doing systematic scientific research into the chemical and physical bases of the properties of materials has become the preferred modus operandi of materials researchers. Call it materials by design: not been there, never done that.

To do materials by design is to dream up new materials with a specific suite of properties, to infer the way you need to put atoms together in order to make your dream materials, and then to follow up the plan by actually making the stuff of your dreams. You might want to make steel twice as strong as the strongest steel that now exists. That way you can use half as much to make each car or skyscraper. You might want to make new medical implant materials that corral the body's own repair mechanisms to regenerate pulverized bone or atrophied muscle. Maybe your aim is to create new semiconductor materials that will be to silicon what silicon became to the vacuum tube.

People have had material dreams like this before, and their dreams have come true, but only by some combination of hit-and-miss experimentation and sheer luck. A new and perhaps ultimate level of human control over the material world is approaching. In this era a materials researcher will be able to tap a description of some never-before-seen material into a computer that has been well stocked with all the relevant corpus of knowledge about the ways of materials. And out of the computer will come the recipe for making the material. In a fully automated CAD/CAM (Computer Aided Design/Computer Aided Manufacturing) setting, the design specification will immediately speed over communications lines to a factory floor where computer-controlled and roboticized procedures will transform starting ingredients into the specified material.

This scenario remains utopian, but not by all that much. There already are a number of cases in which forward-looking materials researchers have pushed the envelope of discovery into this materials-by-design paradigm. There is no reason to think that they are flukes. There is far more reason to believe they are just the first materials designers, akin to the first stone-flakers 2.5 million years ago or the primitive metallurgists who first smelted metal from rock ore 10,000 years ago.

The immensity of a shift to a materials-by-design paradigm is hard to overstate. To date humanity hardly has tapped the diversity of materials latent within the periodic table of the chemical elements. If you begin with sixty or so elements of the table that you might actually consider for making materials, take every possible combination of five different elements in this set, and then vary the proportions of each element in each of these combinations, you already are talking about tens of billions of materials. And that is not even considering the rapid multiplication of this already large number if you consider six-atom or seven-atom combinations, which is where many materials scientists think the most important discoveries lie. Nor does it include that multiplier that comes with countless different ways you might process these compositions using different combinations of temperature, pressure, and other critical manufacturing conditions. The number of materials humanity has so far probed, much less to put to use, probably does not exceed a few million! We have hardly scratched the surface.

What makes materials by design such a heart-quickening prospect is that humanity already has stumbled onto big-time, society-changing materials—stone, bronze, iron, steel, glass, plastic, and silicon—without the general ability to predict what will happen when certain combinations of atoms and molecules are put together according to some particular procedure. And everybody knows there are mother lodes in the periodic table just waiting to be discovered.

Francis DiSalvo, a veteran solid-state chemist at Cornell University, has put it this way: "Every material that can possibly come our way is sitting there in the periodic table." Every diamond, every superconductor, every three-five optoelectronic crystal, every speck of dust, every biological tissue, every material that ever was, is, or will be is there hidden in the periodic table. "It's a puzzle given to us by God, and there has to be a key; maybe there are lots of keys," DiSalvo says. "It's one of the most exciting brain teasers that science will ever have."

Some of those keys, it seems, now are in the hands of a small but growing cadre of visionary materials researchers who have tasted the power of materials by design.

SIX

the RITE OF ATOMIC MASONS

People love to talk. That includes Federico Capasso, a leading materials designer at the part of Bell Laboratories now called Lucent Technologies, who already has found some of Francis DiSalvo's keys to materials essential for the high-capacity communication of today and tomorrow.

Communicating is as basic to human life as breathing and eating. The creation of a telephone network at the turn of the century provided people with a new way to communicate to someone too far away to speak with in person. A telephone call is more intimate and immediate than a letter or even a telegraph message because the two people talking can hear each other's voices. You cannot send laughter in a letter or over a telegraph wire.

By the end of World War II, AT&T and Western Electric (the arm of AT&T that manufactured telephone equipment) had erected a massive national mesh of copper wire through which just about anybody anywhere could reach just about anyone anywhere else. But the nation's hunger for telephonic connection was merely whetted. This was good for the company. But Mervin J. Kelly, then executive vice president at Bell Telephone Laboratories in Murray Hill, New Jersey, where AT&T had thousands of researchers doing the basic and applied research for keeping up with future communications needs, looked at the existing network and got worried. It was the switches, the millions upon millions of switches, that unnerved him.

When a person made a phone call in the late 1940s, a series of little ticking sounds could be heard as a tiny electrical current shot from the caller's phone and then journeyed through a network of electromechanical relays that kept handing off the current until it made it straight into the other person's telephone. An automatic switching station sounded like a roomful of people with chattering teeth. Before automatic switches, it was a roomful of people—operators—that switched the right circuits on and off by plugging and unplugging wires in and out of sockets. In a mechanical switch, or relay, a tiny current would loop around a metal coil to create a magnetic force. The force would move a small metal arm that would open or close a circuit by literally completing or interrupting a conductive pathway for electrons. It was the movements of those arms that made the little ticking sound. Once an unbroken electrical pathway was established between two specific phones, then voices could journey over copper wire back and forth in the form of electrical pulses that could drive a telephone's speaker. What concerned Mervin Kelly and AT&T was all of the ticking going on in the millions upon millions of switches.

They knew that the telephone system would have to expand greatly to meet the country's growing need to communicate. The mechanical switches that were making the miracle of person-to-person, direct-dial phone calls possible would not be able to handle the projected traffic. The switches were too bulky. Their moving parts wore out. And they also needed too much energy to move those little metal arms to make that ticking sound.

Kelly let it be known around Bell Labs that he wanted a replacement for the pillbox-size mechanical relays.[1] Make a switch with as few moving parts as possible—even none, if it were possible. Reduce the amount of energy needed to run each relay. And make them small. That's what management wanted. Electron tubes seemed a ready-made candidate, but they could only work as a switch after heating a metal electron source inside the tubes to white hotness. Moreover, producing them in the quantities required was impractical.

The conundrum inspired Walter Brattain and John Bardeen—whose collective talents included theoretical and experimental physics, chemistry, crystallography, metallurgy, electronics, and precision mechanics—to find a solution.[2] By the end of 1947, they found a way to build all of the functions of a mechanical relay or an electron tube into the invisible movement of electrons inside a crystal. In other words they invented the transistor.

They evaporated a tiny spot of gold onto a crystal of germanium, which had gained some fame among scientists for its ability to detect faint radar signals. When the Bell researchers put a positive charge on the gold spot atop their germanium crystal, they induced a negative charge inside the underlying crystal. That was like clearing a previously blocked pathway for a stronger flow of electrons through the crystal. It became known as a point-contact transistor, a solid-state electronic switch.

The means to meeting Kelly's edict was in hand. The ticking of the metal arm in a mechanical relay could conceivably be replaced by the silent governance of electric charges. There were no moving parts in a transistor, it didn't take much energy to operate, and was potentially very small. For their momentous invention, Brattain, Bardeen, and William Shockley (who invented another type of transistor, the junction transistor, which displaced the point-contact transistor) won a Nobel Prize in 1956.

The Nobel Prize committee had no idea how dramatically the new electronic gadget would change and quicken the world. It was the beginning of the end for big, power-gulping vacuum tubes and the beginning of solid-state electronics. A transistor, after all, was a specially made solid object. It didn't have to be enclosed in glass, so it didn't contain any air or gas of any kind. It didn't involve liquid. It was solid through and through. A decade passed before anyone knew if this seed of solid-state electronics really would germinate and render vacuum tube electronics a has-been technology.

Ten years after the first transistor, Bell researchers invented the laser, and by 1976 a large community of researchers was striving to marry the electronic purity of solid-state devices with the optical purity of laser light. When a young Italian physicist named Federico Capasso arrived at Bell Laboratories in 1976 for what was supposed to have been a nine-month visit, Bell Laboratories had become a Mecca of physics. For Capasso, an optical physicist interested in the many ways that light and solid materials interact, coming to Bell Labs was a dream. Written in metal letters on a wall as you enter AT&T's flagship laboratory in Murray Hill, New Jersey, are these words of Alexander Graham Bell: "Leave the beaten track occasionally and dive into the woods. You will be certain to find something that you have never seen before." They were words that Capasso took to heart. He now heads the Quantum Phenomena and Device Research Department at Lucent Technologies, a portion of the recently reorganized Bell Laboratories.

Capasso, a short, ebullient man with unkempt, thinning hair and a big gap-tooth smile, practices on the leading edge of materials science.[3] He and his colleagues design solid-state materials from atomic scratch, mostly semiconductor crystals in the same general material family as germanium and silicon. These designs rely on the deepest theories of how atoms and electrons and energy behave—quantum mechanics, that is.

Capasso's team uses a miraculous tool called molecular beam epitaxy (MBE) that can spray-paint a crystal into existence one atomic layer at a time. Like the inventors of the transistor, Capasso likes to make tiny solid pieces of crystals that have no moving parts yet behave like machines.

Capasso's work at Bell originally concentrated on the detection of invisible infrared light, which was the region of the spectrum that traveled well in the optical fibers. Nine months passed, and Federico indeed had developed a better feel for the leading edges. But instead of returning to Italy, Federico stayed on at Bell Labs. He never left.

Capasso soon became a pioneer of "band-gap engineering," a term he says he coined himself in 1983 when he and his research kin had acquired the kind of control over matter that the term implies. When Capasso says "band-gap engineering" in his strong Italian accent, it sounds like "bang-up engineering." For now, think of Capasso as an electrician who fiddles not with wires and wall sockets but with the electrons moving inside the tiny atomic jungle gyms within artificial crystals of his own design. Strange things can happen inside these crystals. You can send electrons into them and get pristine beams of laser light to come out of them. You can make them into tiny little computers that can tell you if two numbers are the same or not.

These crystals begin as figments of a prepared imagination. They lie latent between the terms in arcane equations of difficult physical theories. They come from the researchers' intuitions about how electrons move within solid materials. They often are born from "energy diagrams," which are rough sketches that show what the territory within a specific semiconductor crystal might look like to electrons: where within a crystal an electron encounters barriers, repulsions, or open doors. Combining theory, intuition, energy diagrams, and volumes of data from real materials, Capasso and his researchers test and refine their ideas.

Band-gap engineers do not do the kind of physics that

typically catches the public's attention, things like creating a new particle of matter in a gargantuan accelerator or detecting the earliest remnants of the Big Bang with satellite-born instruments. Instead they do solid-state physics. They study how solid materials are built and how they behave, a preoccupation that encompasses a good part of our everyday natural and artificial landscape.

The term *band-gap engineering* is a hieroglyph to most people, even to scientists who have gone through ten years of collegiate training, a postdoc or two, and decades of employment in high-tech settings. Capasso's quick definition doesn't help: "We are using the laws of quantum mechanics to design new materials." Translation: We are using theories and calculations about how electrons behave in crystals to invent new high-tech materials. What is remarkable about this claim is that no one has ever actually been able to say with a straight face that he or she had *designed* a new material. People had always stumbled upon new materials by trial, error, and serendipity, or by varying details of inherited craft knowledge.

The ability to design and make materials with specific properties is, for the materials scientist, tantamount to breaking the 4-minute mile or breaking the sound barrier. It is a celebration of human capability. It is a harbinger of achievements to come. And since it is materials that so often limit existing technologies and spawn new ones, the ability to design materials is a general key to accelerating technological development.

The general goal of materials designers is to uncover previously unknown or unrealized properties that matter can have in principle but has never realized in practice. This approach to materials marks a drastic shift from the past. Since ancient times people have known that the earth harbors various substances that could be useful in their raw states or processed to bring out more useful features. The key to unlocking the multifarious personalities of matter, however, remained hidden well into the nineteenth century. Chemists were learning by trial and error how to build and break molecules, but they had little idea why some things worked and others didn't.

In the 1920s, as scientists were still expanding the vast technological possibilities created by chemically transforming the raw substances of the earth, physicists began to uncover a deeper layer—the quantum mechanical—of the mystery of stuff. Quantum theory began to make sense, albeit strange sense, of why different atoms and molecules behaved the way they do. Quantum theory was a means for the scientific imagination to get far more intimate with atoms and atoms' electrons, the handmaidens of chemistry, so that they could first explain and then predict material behavior. In this way, theorists finally have been able to make discoveries before experimentalists.

Band-gap engineers like Capasso do in fact mine the labyrinthine theoretical framework of quantum mechanics for clues to entirely new materials. Just how they do it requires a bit of elaboration. In the world according to quantum mechanics, the electrons bound to each kind of atom can take on a set of specific energy levels and no levels in between. If an electron in a silicon atom were a car, it could travel at, say, exactly 10 miles per hour, 17 mph, 28, 43, 58, and 81, but never at speeds in between. Give it enough juice at 17 and it goes instantly to 28 without passing through 18, 19, 20, 21, 22, 23, 24, 25, 26, or 27. Strange but true. In a different atom, maybe carbon or titanium or indium, that same electron would have a different characteristic set of energies.

Now snorkel a little deeper into the quantum mechanical abyss. If you pack zillions of atoms into a solid crystal, the rules of electronic energy and behavior can change drastically (electrons are like people; their individual behavior changes when they find themselves in crowds). For one thing, the number of discrete energy levels allowed for an electron bound to one of the crystal's atoms increases to the point where the energy levels smear together into a band of energies—the valence band. At the same time, the crystalline venue brings forth a new, higher band of possible electronic energies—the conduction band—in which electrons are no longer restricted to stay near this or that atom. Instead they can become free agents and flow through the material like water through a pipe. Between the valence band and the conduction band is a no-man's land for electrons, a region of forbidden energies. This is the band gap.

In conductive materials like copper and gold, the conduction band is only half filled, which means that the metal's electrical resistance is small. The slightest impetus, such as hooking these materials into a loop with a tiny battery, will send electrons moving in the conduction band to form an electrical current. This little bit of quantum mechanics happens every time you plug in a lamp or talk on the phone. In insulators like rubber or mica, on the other hand, the gap is so big that getting electrons to jump across would require so much input energy that it would deep-six the material.

Semiconductors, the materials that Capasso spends most of his time thinking about, are an entirely different kind of electronic creature. The most famous of these, silicon, is what Bill Brinkman, vice president of physical science and engineering research at Lucent Technologies, describes as "a material given to you by God." The band gap in a semiconductor is too big for electrons to jump without some external impetus such as a little heat, light, or voltage (which is like a slope for a river of electrons). But it is not so big that the input of energy blasts the material to kingdom come. What's more, each semiconductor material like silicon or germanium or gallium arsenide has a different band gap. Each has its own personality in the arena of electronics and related arenas such as optoelectronics (the CCD detectors in camcorders are optoelectronic devices that turn light stimuli into electronic signals).

Semiconductors with relatively small band gaps make for supersensitive heat and radiation sensors. In the Hubble Space Telescope, it is an array of such small band-gap crystals that gather the dimmest signs of light from the big mirrors.

The body heat of a person 50 yards away from a crystal of mercury cadmium telluride can knock enough electrons into motion within the crystal to set off alarms. The National Security Agency undoubtedly is well versed in the ways of such materials.

Semiconductor materials with somewhat higher band gaps are the stuff of transistors and computer chips. In these materials it takes a little more but not too much more, to get electrons to jump the gap. Silicon does this so well at room temperature that it has become the backbone of a technological revolution. But gallium arsenide has a way with light that silicon doesn't. This is especially true when gallium arsenide is sandwiched between closely related materials with higher band gaps: for example, aluminum gallium arsenide.

This is one of Capasso's favorite topics. Once energized into gallium arsenide's conduction band, electrons can drop back down into the material's valence band by recombining with a positive charge. The flanking materials with high band gaps keep all of the action within the meat of the sandwich. This recombination generates light. Thus do tiny semiconductor lasers send conversations coded as light pulses through glass optical fibers. In every CD player a solid-state laser that works via this recombination principle sends a beam of light over the pattern of pits on the disk and then into another solid state detector that converts the pattern of reflected laser light into the electrical pattern that drives the sound system's speakers.

Silicon and gallium arsenide are but figureheads of a vast population of lesser known semiconductors. By mixing and matching two, three, four, and even more elements from certain columns of the periodic table (the third and fifth, and the second and sixth), you get semiconductor materials with a huge range of band gaps, which means a range of materials with interesting optical and electronic properties. If you can control the way these materials come together in crystals, you open a practical door to these materials.

To solid-state scientists, this mix-and-match game is a technology engine that roars with power. Some "three-five" materials, as they are called, are just right for making superfast transistors, which can be used for speeding up telephone switching systems or computer chips. Others are good for making the infrared lasers that now send light down hundreds of thousands of miles of optical fibers. Others are good for making the yellow- or green-light-emitting diodes that have changed the face of consumer electronic displays. Some are the key to photovoltaic cells for converting sunlight into electricity. In this case the light energy of the sun kicks electrons in the photovoltaic crystal directly into the conduction band. Still other semiconductor crystals can detect radiation. And those are just the more conventional applications. This rain forest of semiconductor diversity has become the favored locale of thousands of materials scientists, physicists, chemists, and electrical engineers, all seeking to discover the stuff that will spawn another Silicon Age.

That opulent diversity notwithstanding, band-gap engineers like Capasso want to squeeze yet more out of the periodic table than is available from straightforward combinations of elements. The specific band gaps you can get out of these atoms when packed into crystals are merely the ones that mother nature has seen fit to bestow. Capasso does not like to be boxed in like that, especially when his designs for new transistors, light detectors, or lasers call for a band gap that isn't in nature's standard catalog.

Soon after Capasso arrived at Bell Labs in 1976, he began to see a way around these seemingly ineluctable natural restraints on band gaps. He sensed that there was an entirely new range of possible band gaps, and he thought he knew how to find them. If he was right, the new freedom to dictate band gaps would lead to a vast new arena of materials, some of which could be as technologically important as silicon.

One of the biggest goads was the same one that led the

managers of research for the Italian Post Office to send Capasso to Bell Labs—the gut belief that communication based on electrical pulses traveling down electrical wires would have to give way to communications systems based on light pulses traveling down some kind of wires for light. Major cities already were running out of underground space for fat copper cables, yet the communications demands of their citizens were accelerating relentlessly. If communicating by light was the wave of the future, researchers would have to devise components that could generate the right kind of light (laser light), guide that light globally, and detect all of the optical commotion.

SPRAY PAINTING WITH ATOMS

In the 1970s the pieces of an optical communications system were emerging from the primordial soup of materials research.

One of the biggest pieces had come from brilliant work at the country's premier glass and ceramics research center at the Corning Glass Company. In 1970 researchers there demonstrated that they could make glass fibers pure enough and leakproof enough that light entering one end would still be detectable many miles down the fiber. It was an astounding achievement, akin to making window glass able to transmit light through miles of thickness. For years this goal had eluded other researchers, who couldn't get rid of enough of the physical defects like tiny bubbles that scatter light in useless directions or chemical impurities that absorb light. They always ended up looking through a glass fiber darkly.

The Corning solution involved evaporating pure silica (the stuff of sand, which is made of silicon and oxygen atoms) into a rod that they then could pull like taffy into astonishingly thin fibers.[4] A crucial and brilliant detail, however, was to ever so slightly contaminate the inner layers of the evaporated silica with other atoms. This resulted in an inner core of glass surrounded by a cladding made of a purer glass formulation. This made the outer cladding a reflective sheath that would send any light hitting it back into the core of the fiber. Try as it might, light in the core just could not leak out.

At last optical fibers could serve as long-distance wires for light. Corning was betting so highly that their fibers would become a mainstay of future communications systems that they built a multimillion-dollar factory in 1976 before they even had a single major order. That put the pressure on places like Bell Labs and Japan's NTT to devise and build the lasers and light detectors that would hook onto either end of these fibers.

Groundwork had been laid already. The same year the Corning researchers made their optical fiber breakthrough, Morton Panish and Izuo Hayashi at Bell Labs had built the kind of solid-state laser that would inject light into these hair-fine conduits of light. By sandwiching a thin layer of gallium arsenide between two layers of aluminum gallium arsenide, they had made a solid-state semiconductor laser that worked at room temperature and that could lase continuously rather than in intermittent spurts. Previous attempts to make such a laser had always resulted in tiny imperfections that allowed the energy in the excited electrons to vent in ways that didn't emit light or that blasted apart the crystal's internal architecture during continuous running of the laser.

Bell was the right place for Capasso for many reasons, but the presence of Alfred Cho was probably the biggest. By 1980, when the two would begin the first of many collaborations, Cho had already spent a dozen years doggedly working out the bugs of an astounding tool called molecular beam epitaxy (MBE). Its potential for growing nearly perfect crystals one atomic layer at a time fueled Capasso's quantum mechanical imagination like kerosene on hot coals.

At the center of an MBE machine is a stainless-steel chamber that looks like it could withstand the pressure at the bottom of the ocean. Inside there are thick glass portholes ringed with heavy-duty bolts like submarine hatches. Tubes go into the chamber at every angle, and there are mysterious appendages and boxes all over. Jumbles of wires, which seem to extend in all directions, terminate in racks of electronic boxes and a control station. The machine looks like a metallic Medusa. Separated by shutters from the main chamber are small furnaces with crucibles containing the atomic ingredients needed to make a band-gap engineer's dreams come true. When the crucibles are heated, such atoms as silicon, arsenic, gallium, aluminum, indium, and phosphorus jerk loose and start to diffuse into the MBE chamber. (See Figure 21 on page 25.)

By controlling the shutters with a computer, an MBE operator directs beams of various atoms into the growth chamber and onto the surface of a crystal template waiting there like an unpainted wall—except that this wall is a crystal surface resembling an endless atomic-scale egg carton. Each barrage of atoms coming from one or more of the crucibles fills up the corrugations in the crystal surface to produce a brand-new crystal layer ready to receive the next barrage. The template crystal is heated to ensure that the incoming atoms will scramble around until they find a site at which the chemical attraction is strong enough to hold it. Cho himself likens MBE to spray painting with atoms.

The process is inherently slow, though researchers have found ways of speeding it up. It typically takes about one hour to grow one high-quality micron's worth of a crystal. It would take a few days to get a crystal as thick as a sheet of paper. In a year the crystal still wouldn't span half an inch. A bonus of this sluggishness, however, is that MBE operators can decide just how many layers of each element or elemental combination they want. Moreover, as they go, they can leak tiny amounts of atoms known as dopants into the beams. These can provide the electronic or optical spice that can make these constructions scientifically or technologically interesting.

To visualize what happens in an MBE machine, think of selling eighty thousand general admission tickets to the Super Bowl and then just opening the gates an hour before game time.

Even if by some miracle the rushing crowd managed to fill all of the seats in perfect order, that would only represent a tiny fraction of what needs to happen to grow a crystal. Imagine now if the stadium had billions upon billions of seats extending indefinitely in all horizontal directions and stacked hundreds or thousands of layers high. For MBE to work well enough for the crystal-building purposes envisioned by Capasso, atoms leaked into the vacuum chamber would have to completely fill that enormous three-dimensional stadium almost flawlessly.

Lucky for Cho, atoms are not quite as unruly or as unpredictable as rowdy football fans. Over the years he reduced temperature fluctuations in the instrument, got rid of impurities from the ingredients and machine, and disarmed other gremlins that would pock the resulting crystals with too many defects. The technique was so good by the mid-1980s that it became the manufacturing tool of choice for making most of the semiconductor lasers shining invisibly inside compact disc players. "Over 70 percent of the world supply of compact disc lasers are made by MBE," Cho likes to point out. MBE is now so good that a square centimeter of crystalline real estate—an enormous expanse of over 1 quadrillion atoms—can have as few as three atoms out of place. That is like searching through a population one hundred thousand times that of the earth before finding a crazy person. Solid-state devices of the not-too-distant future may require that kind of material perfection.

So there was Capasso in the late 1970s within a stone's throw of perhaps the only machine in the world capable of building a crystal atomic layer by atomic layer. "It was knowing that MBE was there that made me really think along the direction of band-gap engineering," Capasso recalls. The kind of unprecedented control over matter that MBE could provide, Capasso intuited, could prove to be the key for turning a lot of quantum mechanical dreaming into real, hold-in-your-hand embodiments. MBE was a tool that could turn equations, physical intuition, and back-of-the-envelope energy diagrams into reality. Moreover, it was just this kind of thought-to-stuff transformation that eventually helped to take Bell Labs and others to the threshold of the era of light-based communications.

THE ROAD TO BAND-GAP ENGINEERING

Until Cho and Capasso began collaborating in 1984, no one had ever deliberately done the kind of fundamental fiddling that band-gap engineers do. People had thought about it, done calculations, and made predictions. Now MBE was ready to put all of this effort to the test. "The 1980s were for solid-state science a golden era," Capasso says.

To be sure, others had begun paving the way to these good times. In the early 1950s William Shockley had an inkling of the technological gold to come. He surmised that a semiconductor crystal made out of two different materials with an atomically abrupt interface—which no one knew how to make then—would lead to higher-performance transistors (electronic switches) than would be possible with germanium or silicon alone.

In 1957 Harry Kroemer, then at RCA (whose corporate empire was built on vacuum tubes), took another important conceptual step toward band-gap engineering. Kroemer's idea was that by slowly changing the composition of a semiconductor crystal, you could build what amounts to an electron propulsion system into the crystal itself. Electrons wouldn't just mosey on through—they would rush forward as though they had entered a river's rapids. It took thirty years before anyone put ideas like that to the test.

A few years later, Kroemer, whom Capasso considers an "intellectual father," had another idea. He suggested that sandwiches of certain semiconductor materials would actually behave as tiny solid-state lasers. At the time, lasers, which were first proposed theoretically in 1958, were bulky affairs in which the light emerged from excited electrons in gases. The first working laser was built two years later. Laser light is the result of electrons getting pumped up to a higher energy state and then falling back down to earth with a blaze of one-color light. Normal light sources emit many colors or wavelengths, which is why they appear white.

In 1970 IBM—then one of Bell Lab's few rivals in research and development—got into the act. Leo Esaki (another elder of band-gap engineering who later won a Nobel Prize) and Raphael Tsu, then both at the IBM Thomas J. Watson Research Center in Yorktown Heights, New York, published a theoretical paper in IBM's house research journal.[5] It became one of the most widely cited sources for a semiconductor structure called a quantum well, which became for Capasso and other band-gap engineers what two-by-fours are to carpenters. (See Figure 32.)

Like water wells or inkwells, quantum wells are good at confining something. Unlike wells that people can fall into or that can hold ink, quantum wells are measured in angstroms (a ten billionth of a meter). They are sometimes only a few atomic layers thick. Under a souped-up microscope, a quantum well looks like a crystalline sandwich. The meat is a layer of one material like gallium arsenide, and the bread is made of another material like gallium aluminum arsenide. These kinds of wells confine electrons to discrete energy levels much the way a string fastened between two fixed pegs is constrained to vibrate

---

Atomic Construction. Each blob in this electron microscope image represents an atom of either indium, gallium, arsenic, or phosphorus. The more tightly packed group of three layers in the middle is indium gallium arsenide, flanked on top and bottom by thicker layers of indium phosphide. At the boundaries of the different materials there is an abrupt change in the potential energy. To an electron passing through this three-layer molecular sandwich, these potential energy changes, or steps, amount to an electron launcher. Such launchers, in turn, can serve as building blocks for solid-state devices such as radiation detectors or photomultipliers that depend upon cascades of accelerated electrons. This specific structure was grown by Morton Panish at Bell Laboratories using molecular beam epitaxy.

BELL LABORATORIES, COURTESY J M GIBSON AND S N G CHU

FIGURE 32

only at discrete frequencies. Some of the electrons in the current moving though the crystal fall into these quantum wells. Once in, they have a hard time climbing out. That captivity makes them a little crazy.

A year after Esaki and Tsu floated their theoretical idea, two Soviet physicists then working in Moscow, Rudy Kazarinov (who ended up working in Capasso's group) and Robert Suris, had registered another key idea in an obscure Soviet journal on semiconductors.[6] They wrote that quantum wells, if arranged in a series, could be a route to a new type of solid-state laser. The wavelength or color emitted by the laser could actually be chosen—dialed in, so to speak. Electrons would drop—or "tunnel," in the parlance of quantum mechanics—from quantum well to quantum well, losing energy each time by emitting photons of wavelengths that correspond to the energy jump. Almost a quarter century later, this became a central design principle for Capasso's most prized achievement. If these Russians were right, what seemed like an arcane theoretical suggestion by the researchers at IBM could be the fount of some very practical technologies relevant to optical communication.

Capasso's first real foray into band-gap engineering began with a project to make light detectors, a key component for optical communication. He was still working at AT&T's facility in Holmdel, New Jersey, which was fated to become the research and administrative headquarters for the company's optical communications business, including its massive business crisscrossing the oceans with undersea optical cables.

Capasso took on the challenge of trying to develop a new breed of light detectors that would work well in the specific wavelengths that traveled best down optical fibers, namely, 1.3 to 1.55 microns, which are invisible infrared wavelengths. The most sensitive kind of light detector available was known as an avalanche photodiode (APD) and was made out of the old standard—silicon.

In an avalanche photodiode, a process known as impact

ionization amplifies tiny light signals into large electronic signals. The incoming light knocks loose a few electrons, which then gain energy from the high voltage applied to the device. For the bumped electrons, this is like falling off a riverbank into the raging current. These energized electrons then knock more electrons from the silicon's valence band into the conduction band. The high voltage, however, also throttles up some of the newly knocked electrons to knock yet more electrons from the valence band to the conduction band, and so on, until you have an avalanche of electrons careening into a metal lead. A tiny light signal going into the gadget comes out as a detectable electronic signal.

That is just the sort of detector everybody wanted in a long-distance optical fiber system in which the light at the end of the glass fiber would be a mere ghost of what went in. Otherwise, the system would have to be so riddled with "repeaters"—expensive and complicated components that reconstitute and amplify the weakened light signals for the next leg of the journey—that prospects such as undersea fiber-optic cables would be impracticable.

APDs were clearly needed, but making them of silicon was problematic. The wavelength of light that triggers an electronic avalanche within its crystal flesh is not the wavelength that travels best down optical fiber. An optical fiber rigged with a silicon detector is like a music lover wearing earmuffs at a concert.

"The materials had to catch up to this minimum wavelength of 1.3 microns," Capasso recalls. Researchers, knowing that new semiconductor materials were in order, tried all kinds, including some with three and four different atomic ingredients. What eventually emerged as the standard materials for APDs, which are used in long-distance communications systems, is a combination of indium phosphide and indium gallium arsenide. But this combination had its own failings for the higher rate of data transmission needed for an information highway without major traffic jams: it detects 1.3 micron light but produces a noisy electronic signal that limits the rate at which it can reliably detect data. So for higher-capacity communications, the standard APDs would work only with stronger light signals that could emerge like a distinctive voice through the chatter of a cocktail party. The requirements were clear: either very strong lasers or an unwieldy population of repeaters that could doom, for example, a transoceanic application.

"The materials that people were using to make avalanche photodiodes for light communication at 1.3 to 1.55 microns didn't have low enough noise," Capasso recalls. The noise comes from the fact that electrons are not the only things that get sent into motion when light hits these detectors. Positive charges, which are called holes because they are the same as the absence of an electron, end up streaming in the other direction. When they do this, they too knock other electrons loose, which in turn do the same. This is counterproductive feedback and ends up reducing the sensitivity of the device while making it harder for the real signal to shine through.

It would take some very different thinking to get rid of the noisy feedback. As a materials designer, Capasso tended to work backwards, to imagine the properties he wanted in a material and then to figure out the atomic arrangement that might actually perform that way. In essence, he dreamed up materials and then convinced a crystal grower to make his dreams come true. It had always been the other way around: someone would make a material that happened to have specific properties, and then people would try to figure out how to put the new stuff to use.

For the avalanche photodiode challenge, Capasso reasoned that in the ideal material, electrons would produce an avalanche of other charges but that the positively charged holes would not. He knew that such a material did not exist and that it probably shouldn't even be able to exist. The dogma was that if electrons get knocked into motion in a detector, holes will be sent into motion, too. Nevertheless, he persisted. "I asked myself, can I create an artificial material where only one type of charge carrier creates an avalanche?" Capasso recalls.

It is amidst challenges of this sort that intuition saves the day. Capasso says, "There was no law that said it could not be done, and usually in nature where there is something that is not forbidden, you know that eventually it can happen." This could be a slogan for band-gap engineers. The technological carrot was there, too, according to Capasso: "I knew that if I could come up with such a material, people would use it in lightwave communication."

In the summer of 1978, it all began to coalesce in Capasso's mind. "All of a sudden one day I got the idea. We had to make what essentially would be a launching ramp or a ski ramp for electrons." In such a material any electron knocked loose by a photon would take off like a bat out of hell, kicking up electrons and holes in its wake while the hole would stir up far less electronic activity. He even had an idea of how to realize this situation—creating a structure of alternating layers of aluminum arsenide and aluminum gallium arsenide. This material striping was similar to the superlattices that Esaki and Tsu had suggested years earlier.

Gallium arsenide and aluminum gallium arsenide were natural choices. For one thing, they were familiar to researchers. For another, their quantum mechanical properties (such as band gaps) were quite suitable for creating the ski ramp that Capasso had in mind. The abrupt transitions between the two alternating materials would create energy steps, which are larger for electrons than for holes. That means that the electrons would gain more energy than holes and would be more effective at creating electron avalanches then hole avalanches.

As soon as he floated the idea, doubters sprang to the fore. Some said the electrons wouldn't be able to make it up an energy barrier in front of the ski ramp that Capasso could see was necessary from the energy diagrams describing the low-noise APD. And a stranded electron is a dead end for an avalanche photodiode. There were a lot of shaking heads.

All of that nay-saying only spurred Capasso on. "I was so upset that I said, 'I am going to solve this, and that is that.'" He combined what he knew and intuited about band gaps with what he could calculate about how electrons behave in solids. Then he designed a superlattice—a stripe of gallium arsenide, a stripe of aluminum gallium arsenide, a stripe of gallium arsenide, and so on. Each layer in the blueprint had a slightly different thickness. Calculations showed that such a nuanced architecture would lead to a crystal harboring a series of regions with different band gaps that collectively would produce ski ramps for an electron.

At this point in a Capasso project, a technical relay sets into motion. The blueprint for the crystal goes to the MBE lab, where a wafer of the designed crystal is grown. The wafer goes to another colleague, who dices out tiny pieces of crystalline real estate and attaches metal leads to the little plots. The specimens are then ready for physical testing to see what they can do.

The experiment to see if light triggers an avalanche of electrons—but not an avalanche of holes—in a piece of crystal is to inject a specific dose of light on one side and then see if an inordinate amount of electrons comes out the other. The input-output ratio should be akin to whispering into a mike and then hearing a thunderous sound come out of the speakers. In Capasso's design there also should be a lot less noise from holes.

In the 1980s Capasso and various colleagues indeed came up with a family of APDs, many of which performed just as predicted. These researchers actually had dreamed up materials with capabilities never before seen and then made them real. But it was a somewhat anticlimactic experience. The problem was that the growth process was too complicated and inefficient for commerical production. Although Capasso rates that experience as "one of the greatest satisfactions" of his life, no one else paid much attention.

Starting in the mid-1980s, however, people did begin taking notice. Eventually some of them, Japanese researchers mostly, "jumped all over it," according to Capasso. Researchers at three Japanese companies—Hitachi, NEC, and NTT—began work on multiquantum-well APDs that were similar to Capasso's or based on the same principle. A decade after Capasso's team had first introduced them, the devices finally were getting close to commercial viability; now AT&T also has begun developing them.

All of that Sturm und Drang over whether his ski-ramp idea would really work led Capasso to another insight. Applying a voltage across a superlattice changes the energy landscape within the crystal. He realized that he could actually transform the energy landscape from one resembling a series of walls and wells into something that looked like a staircase. Says Capasso, "This would be an artificial material, which could be used to make the solid-state equivalent of the photomultiplier tube." It is a goal much like Mervin Kelly's call, forty years earlier, for a solid equivalent of the mechanical relays that were running the phone network in the 1940s. A PMT, as a photomultiplier tube is called, is a tube that multiplies light energy so that you can detect little bits of it or use less radiation to achieve goals like taking medical X rays. It multiplies electrons like an avalanche photodiode, but there are no holes, and so it is less noisy. Its drawback is that it operates under much higher voltages. (See Figure 33.)

That kind of signal amplification in an MBE-made crystal is just the kind of material talent that makes for world-class radiation detectors. By 1995 neither Capasso nor anyone else had actually realized a solid-state PMT. But the fact that Capasso could actually come up with a blueprint for such a thing was a reminder that he had found a door—band-gap engineering—to a mother lode of scientific and technological leads.

In the early 1980s, with exquisite control over semiconductor crystal structure, with quantum mechanics as a fountain of leads, and with the overall technology goals of Bell Labs as a context, Capasso began to perform thought experiments, mental simulations of what it is like to be an electron inside of a crystal: "I am an electron sitting in this quantum well. . . ." He was developing an intuition for what would happen in different superlattices harboring different landscapes of band gaps.

By the end of the decade, Capasso was giving talks at meetings about ever more exotic devices grown by MBE, not just detectors and lasers. One of them was the resonant tunneling transistors in which electrons tunnel between wells at a handful of specific energy levels. This opened a route to a new kind of microelectronic circuitry whose logic was more powerful than the simple on–off, yes–no, 1–0 binary logic underlying conventional electronic circuitry based on standard transistors.

Back-of-the-Envelope Pictures of Energy Inside a Crystal. Band-gap engineers like Federico Capasso rely heavily on simplified portrayals of how the energy landscape inside a complicated crystal changes at different locations within the crystal. Top: Each upwardly rising stage of this energy diagram corresponds to many layers of a crystal grown so that the amount of energy needed for an electron to jump from being bound to an atom to being free to conduct continually increases. Putting several similarly grown stages in sequence leads to an energy diagram that looks like a sawtooth. In this case, an electron continuously gains energy and then falls down the end of a tooth at which point more energy must be injected to get the electron to travel up the slope of the next sawtooth. Bottom: When the same crystal is sandwiched between electric contacts and a voltage is applied, the electric field tilts the sawtooth landscape downward into a staircase-like situation. A free electron created by absorption of light within the crystal goes down the first step gaining kinetic energy. Some of this energy can transfer to some more of the crystal's electrons, which then begin to move down the next energy step all the while gaining kinetic energy. This process is repeated at each step, leading to a multiplication of the initial electrons. The result is a crystal that behaves like a solid-state photomultiplier: it responds to incoming photons, or light, with a disproportionately large electric signal.
BELL LABORATORIES/LUCENT TECHNOLOGIES

FIGURE 33

Staircase Solid-State Photomultiplier

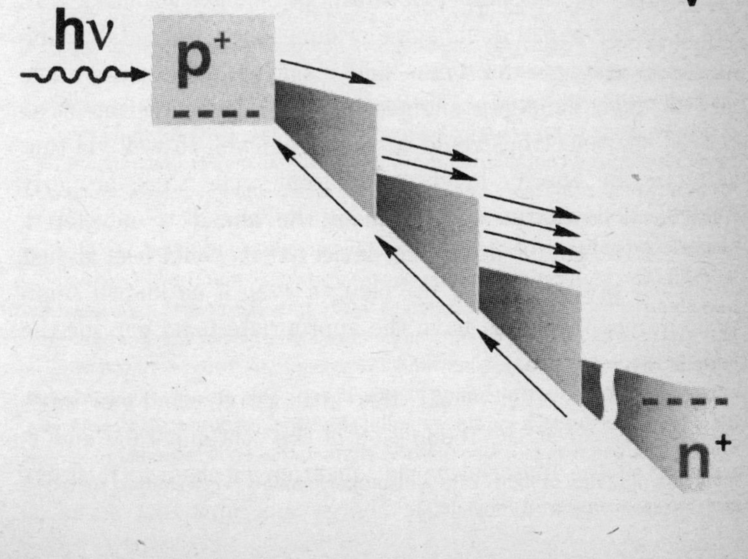

Each of the tunneling energies of resonant tunneling devices could serve as a definable electronic state, each one a value for a multivalue logic. So each one of these quantum devices presumably could do the role of many standard transistors.

It was a nice idea, but using multivalued logic and circuit components was too radical a departure to find acceptance in any but the most specialized applications, at least for the moment. Actually using these devices entails a new kind of programming, a raft of new ancillary electronic components that can handle the multistate logic, and plenty of other unconventional requirements and headaches. Capasso did prove the principle by using resonant tunneling devices to make a parity checker, which looks for differences between binary numbers; this is essentially what a computer does when it checks to see if two stored documents are the same or if they have some differences. It takes several dozen conventional electronic components to do the same job.

These dream-to-hardware feats were just warm-up exercises. They were good for training and for demonstrating that the audacious ambition of designing materials from atomic scratch was not the empty optimism of deluded visionaries. In 1986 Capasso began doing some preliminary work and drawing up energy diagrams for a new kind of solid-state laser realizable only through band-gap engineering. In a key experiment he showed that electrons could cascade from well to well via tunneling if the energy levels of adjacent wells were properly aligned. Following this experiment he aimed to develop a generic strategy for designing devices that would lase at just about any predetermined wavelength even if no known semiconductor happened to have the appropriate band gap for the desired wavelength.

Like the earlier work, this attempt at band-gap engineering started at the foundation of the contemporary understanding of the material world: quantum mechanics. It starts with equations and the little energy diagrams that serve as architectural floor plans. These diagrams show rather simply just where electrons can or cannot go and how hard it is to push them from one place to another within a complicated semiconductor crystal. When he is at work, Capasso probably does not go an hour without scribbling down an energy diagram into a notebook or blackboard or the back of an envelope.

For Capasso a picture is worth a thousand equations, and the pictures seemed to be telling him that a new king of laser, the quantum cascade laser (QCL), could work. In conventional semiconductor lasers (like the ones found in compact disc players) the laser light comes from excited electrons when they swan dive across the material's band gap into the valence band, where they recombine with holes. The energy difference between the excited and ground state is equivalent to the energy of the emitted photon, which means that the wavelength of light is determined by the material's band gap. Normally, the gap is itself completely determined by the material's chemical composition. That is where the QCL idea differed so radically. The presumed physics of light production did not involve holes. Instead, electrons alone would produce the light as a by-product of the pathway they would be forced to take as they traveled through the energy landscape of a precisely constructed semiconductor crystal. In the QCL, electrons are constrained to make transitions between the discrete energy levels of quantum wells. So the photons the electrons emit have energies equivalent to the energy difference between the wells. The key here is that the energy difference between wells strongly depends upon how thick the wells are, a factor that an MBE operator can dial in.

The bottom line of all of this is that the wavelengths emitted by the electrons therefore can be tuned over a wide range using the same combinations of materials but with different well thicknesses. So the chemical composition of the laser no longer limits the wavelength to a specific value, as had always been the case. To build an actual QCL would amount to an end run around what had been a Berlin Wall in solid-state laser research: a material's intrinsic band gap no longer would determine the actual wavelength that would come out of a laser made of that material.

The blueprint for the quantum cascade laser depicts a through-the-looking-glass skyscraper—five hundred crystalline stories tall yet shorter than a mite's eye. The finished product would be no more than a few micrometers thick, or about the diameter of a bacterium.

It was Barbara Sivco who got the task of actually growing the beast. She had been working for Alfred Cho since 1982. Before that she had grown semiconductor crystals, including gallium arsenide, by more conventional means not so different from growing ice by letting water freeze. But in 1994 she got a request that she could hardly believe from Jerome Faist, a post-doctoral associate from IBM's premier research facility in Switzerland whom Capasso had recruited.

Like Capasso, Faist was a quantum mechanic who got his Ph.D. in laser physics and who always wanted to work at the famous Bell Laboratories. In 1991 Capasso hired him to do the calculations and the hard quantum mechanical thinking that might lead to the QCL, which at the time was little more than an energy diagram. Faist shared an office with Carlo Sirtori, another quantum mechanics aficionado, and the two worked like a tag team whose charge was to come up with the detailed blue-prints for the new species of laser.

"We started out on the wrong track," Faist recalls. Their first idea was to use a standard laser to pump electrons to the top of the energy landscape within their theoretical quantum cascade laser; physicists often use one laser to stimulate another. But that tack restricted the types of crystal structures that the researchers could play with. Even if it worked, it would prove to be an engineering headache later. Another laser requires a lot more space and a rig of carefully focused and aligned mirrors and lenses. It would be much more convenient to power a solid-state laser with electricity than with light from a second laser. After a year and a half and little to show for it, not even a paper, Faist was getting a little nervous. Like Capasso's, his original contract at Bell was for a temporary position.

The smell of failure led him to make a fundamental measurement that would tell him for sure whether the quantum cascade laser was a real possibility. He had to measure how long an electron could remain in an excited state within a pared-down predecessor to the QCL. "There was some danger that the electron lifetime would be too short," Faist recalls. In that case the game would be over before it started because the electron would vent its energy in any number of ways within the crystal before it spit out any light.

To make the measurement he came up with a relatively simple structure. It was a series of nine quantum wells that his calculations suggested should emit dim though measurable light if the lifetimes of the excited electrons were OK. At it turns out, his measurement agreed closely with what earlier theoretical calculations had predicted would be the lifetime for these excited electrons within a QCL structure. Faist recalls, "It was an enormous intellectual boost because it gave me confidence in the predictive power of the theory."

Now he had to grow a structure that actually emitted laser light from nothing other than electrons clunk-clunk-clunking down an energy staircase. As he began to do more detailed microarchitectural designing, Faist quickly realized just how complicated the crystal skyscraper would have to be. "The first time I showed Federico what I thought we would have to do, he had a gut feeling that it wouldn't work," Faist recalls, though Capasso remembers being more upbeat. Faist's appointment was scheduled to end soon, and he began to wonder if he, his wife, and two boys would have to pack up and return to Switzerland with that disappointing feeling that comes when big investments go bust.

When he first gave the QCL plans to Sivco in late January 1994, she almost laughed. She had grown hundreds and hundreds of crystals using MBE for a decade, but no one had ever asked her to grow five hundred layers of anything. But it took her only an hour to program the computer and then about three hours to grow the entire microskyscraper. For a state-of-the-art MBE, five hundred layers is a morning's work. She scheduled the run for a Friday.

Into the chamber Sivco placed the foundation: a microslab of indium phosphide. Then came the three hours of gentle clicking of computer-controlled shutters opening and closing. Hot atoms of aluminum, indium, arsenic, and gallium streamed from the shutters, silently building into layer after layer of differently composed crystal planes. Each layer had its own specified chemical composition and width. Each layer would perform a particular role in a finished QCL like a gear or a pinion or a gasket in a complicated mechanical machine.

The atomic masonry of the quantum cascade laser is intricate indeed. In it is a quantum mechanical staircase of twenty-five steps, each one made of a triplet of quantum wells whose relative proportions of aluminum, gallium, indium, and arsenic atoms are slightly altered to make sure that the electrons continue their downward trek. This is like steepening the downward slope of an electronic river. If these layers actually worked according to the plan, the electrons would emit light as they tunneled downward from step to step.

In this case the color would be invisible. The first QCL Capasso's team designed was for an infrared light source, a fact that would make the lasers good candidates for technologies that can readily see through the atmosphere and monitor pollutants or weather conditions.

Even after these twenty-five trios of quantum wells, the blueprint for the QCL called for another few microns of layer-by-layer construction. Finally, the whole business would get a 20-nanometer-thick cap of gallium indium arsenide sprinkled with tin atoms to provide a final conductive pathway for electrons to reach a metallic overlay. That overlay is what would allow the researchers to hook the QCL to the real world.

If all the electrons in a QCL did was hop down an energy stairwell, few people would have been so interested. Like people who hop and rub their tummies simultaneously, electrons in a QCL are supposed to bound down stairs while simultaneously emitting photons of specific wavelengths. When Faist handed Sivco the plans, no one knew if the new material would be able to pull off the double stunt.

The fateful Friday arrived. Sivco prepped the MBE machine and did the QCL run that had been three years in the making. What came out was a brittle little wafer the size of a quarter. Its shiny, dark, mirror-smooth surface gave it the look of a single undersized lens for sunglasses. (See Figure 34.)

Sivco's MBE performance using Cho's meticulously raised metallic Medusa is the equivalent of a Yitzhak Perlman rendition of a baroque concerto, replete with presto bursts packing more perfectly executed notes per unit time than seems possible. The slightest irregularities—in the pumps that maintain the vacuum or in the temperature of the indium phosphide substrate—the slightest vibrations, or any one of many unnoticeable disturbances could lead to structural imperfections even at the three hundred eighty-seventh layer. The tiniest glitch might prove fatal. Not until Faist got his hands on the crystal would anyone know if the run was a success.

Al Hutchinson, another longtime collaborator of Capasso's who specializes in prepping samples for experiments, got the wafer from Sivco and prepared test specimens for Faist. Executing a number of jewelerlike procedures, he cut tiny stripes from the wafer, etched away a protective overlay without removing any underlying layers, and attached metal leads so that experiments could be done. Faist asked Hutchinson to prepare only small stripes of material from the wafer Sivco gave him. He knew that the larger the stripe, the higher the likelihood of

FIGURE 34

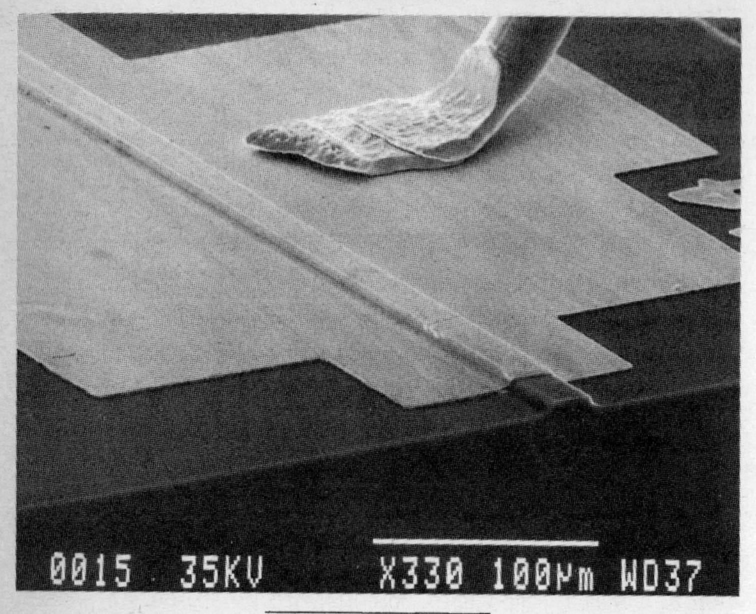

The Quantum Cascade Layer Gallery. To the eye the crystal sliver harboring a quantum cascade laser could just as well be a sliver of anything dark and shiny. Under a scanning electron microscope, however, the mesalike structure of a QCL overlaid with a gold electric contact layer can be seen along with a wire leading to a rack of electronics (above). The mesa is about 25 microns across, which is finer than a human hair. A much higher-resolution cross-section of a quantum cascade laser structure (in the mesa) seen in a transmission electron micrograph reveals a handful of its hundreds of molecular layers (top right). Distinct collections of layers in the structure serve different functions, including accelerating electrons to higher energies and generating light as electrons race down an energy staircase, emitting photons all the way. The energy diagrams depicts some of the physics going on in these multilayered structures (bottom right). In this version of a QCL, which operates at room temperature, electrons enter a "funnel injector" that ramps the electron's energy up so that it enters the top of an energy step. When the electron then falls down to the next step, light is emitted. At the bottom of the next stair, the electron then enters another injector region that then places the electron at the top of the next energy step, from which it falls and emits light—and so on. By adjusting the widths of the various layers of the crystal, the researchers can dial in the wavelength of laser light that comes out as the electron goes down the energy steps.
BELL LABORATORIES/LUCENT TECHNOLOGIES

FIGURE 35

MICROGRAPH OF CROSS-SECTION OF QUANTUM CASCADE LASER BY TRANSMISSION ELECTRON MICROSCOPY

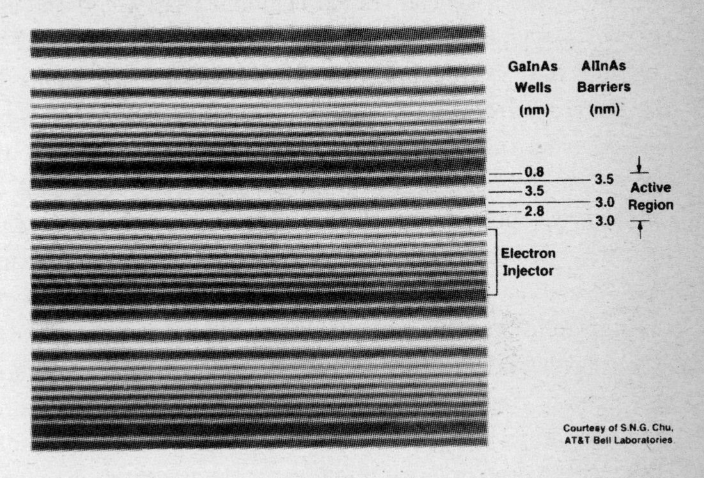

FIGURE 36

ROOM TEMPERATURE QUANTUM CASCADE LASER

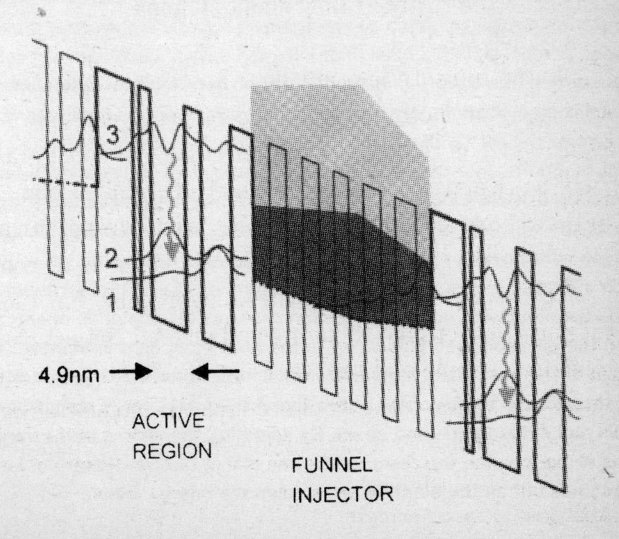

light-eating defects. They decided to go for stripes 2 millimeters long and 10 microns wide, a barely visible splinter.

Faist got the first prepped stripe in the early afternoon. He put it into a special chilled chamber that would enable him to cool the splinter to a few degrees above absolute zero, enabling him to freeze out atomic vibrations that could divert the electrons from their intended action. He connected contacts coming from the frigid crystal to an electronic box that would allow him to control the voltage and the current. A light detector was poised on the other side of the stripe to detect and measure whatever light might come out. For a laser the amount of light coming out should increase as the amount of electrical current going in increases. That's just like an ordinary lamp. If it is a laser and not just a lamp, however, there will be some threshold of current at which there is an abrupt increase in the light output, as though someone suddenly flicked on a floodlight. Faist had that anxious feeling of being on a starting line waiting for the sound of the starting gun.

Ready, set, BANG!

Nothing happened. No matter how far he cranked the current, no light came out of the stripe. Faist asked Hutchinson to get another stripe ready. It took about an hour.

Ready, set, BANG!

Again, nothing. In another hour another stripe was ready for testing. Again, nothing. It had been three years since Faist had begun work on the QCL, and there now was some reason to fear that he, Capasso, and Sirtori had overlooked some fatal flaw in the theory. Or maybe something went wrong during the growth of the crystal. Perhaps they would have to convince Sivco to grow another five-hundred-layer cake for them. A hopeful mind blocks such thoughts and instead conjectures that the problem is simple and fixable. Faist decided to try a shorter splinter of the wafer. That would reduce the amount of light he could expect, but it also would cut down on the probability that defects would spoil the day.

Friday had become Friday night. Faist was thinking about going home to his wife and two kids. Then he began cranking up the current on the small splinter while watching to see if the detector was discerning any light. This time he saw a signal! Light was coming out of that tiny splinter. But was it laser light? At the moment he could only say that he had made a tiny sophisticated lamp, not a newfangled solid-state laser. "Then I saw that the light signal went up and up and up. I knew then that either the device was burning or lasing," Faist recounts. "I couldn't believe my eyes."

He was alone in the lab and so couldn't immediately share the elation of the moment. So he phoned Capasso with the good news. Capasso, who already has enough enthusiasm to perk up a funeral, was gushing. A colleague said she imagined Capasso jumping up and down screaming "It's a laser!" like Dr. Frankenstein. Fearing that if he blinked the success would disappear, Faist took data well into the night.

The next day he and Capasso met at the lab and took more data. "There was no doubt about it," Faist recalls. The little intuition that became a detailed set of calculations that became a blueprint for a five-hundred-layer crystal no thicker than a sperm cell that was made real by MBE actually confirmed the little intuition—another scientist's dream come true. "That was a great way to blow the weekend," Faist says. He, Capasso, and the rest of the team had pulled off an audacious act of materials design.

A week later Dave Lang, Capasso's immediate supervisor, arranged for Bill Brinkman and Arno Penzias, the vice president and president of research at Bell Labs, respectively, to come into the lab and see the laser in action. Penzias had won a Nobel Prize for discovering the everywhere-the-same background radiation that cosmologists had predicted would be present if the universe did in fact begin in a big bang. The triumphant band-gap engineers were still shocked enough at their own success that they considered simply showing data they had collected earlier rather than flicking the switch again and risking an embarrassing moment of nonlasing. But nothing bad happened. The little laser that could lased for the supervisors as it had for Faist and Capasso.

Brinkman knew that the steep peak on the oscilloscope meant he was witnessing a laser like no others. "We had talked about making a laser like this for years," Brinkman notes, recalling earlier discussions with the researchers. "And there it was. This was a real tour de force. This is a step toward the ultimate situation in which you sit at a computer, design a material, and then let a machine make it for you."

The team wasted little time in shooting a paper about the QCL feat to *Science* magazine. *Science*, in turn, wasted little time in publishing it. The paper arrived on February 14 and appeared in the April 22, 1994, edition.[7] It is a rare paper that makes it through the gauntlet of peer review and publication that quickly.

The scope of the list of potential applications increases as Capasso's team improves the QCL lasers and makes them more convenient to use. For one thing, the researchers have gotten these lasers to work at higher temperatures than in their original experiments. In particular, they now can work at temperatures readily maintained with liquid nitrogen, which anyone can carry around in an insulated container. Some designs of the laser even work in sauna temperatures, though the power output diminishes. The QCL team also has become adept at changing the thicknesses of the layers in the active lasing regions of QCLs and has proved that the resulting wavelengths of light that come out can be tailored over a wide range, from 3 microns to 13 microns.

This range is suitable for a smorgasbord of potential applications. High on the list are QCL-based systems for monitoring atmospheric pollution, since the lasers' wavelengths are not absorbed by moisture and aerosols. Another potential use is in industrial process control, since many nasty compounds absorb the wavelength of infrared light emitted by the lasers. There is a bounty of other QCL applications: central components in systems for detecting illegal drugs and explosives; tools that diagnose automotive combustion and emission; collision-avoidance radars in fleets of "smart cars"; military countermeasures; medical testing; and basic research.

The quantum cascade laser is one of the most sophisticated, future-is-now examples of materials design. Charles Townes, a Nobel laureate for the invention of the laser, said of the QCL work that "it opens the door to very important new laser possibilities." It is an outgrowth of the postwar ambition to create complicated machines in single pieces of solid materials by precisely controlling the materials' internal structure. The ability to design materials in this way is a harbinger of the new human relationship to matter.

Quantum cascade lasers occupy a leading edge of current technology. There is an elegance about them, with their pristine crystallinity and their output of pure laser light. They are the lacework of the material world whose mass is measured in tiny fractions of grams. But steel, one of the most traditional materials produced globally at the annual clip of nearly 1 billion metric tons, also has begun to enter the growing club of materials that can be designed from atomic scratch. It seems that steel—the secret of the Samurai, the heart of mythical Excalibur, the skeleton of urban skyscapes, and the very stuff of the industrial revolution—has only begun to show its mettle.

SEVEN

COMPOSING *Steel*

Steel. Silver-gray, cold, solid, hard, strong. What else but steel could have become the metaphor of Superman's superhuman strength? Steel's strength opened the way for a word like *skyscraper.* Without steel the urban skylines would still look like the pyramid-studded Valley of Kings in the Nile. Before steel the only way to build high was to pile stone upon stone. For a hundred years architects and engineers have been relying on skeletons of steel to create the high-rise cityscapes that are among the most recognized symbols of high-tech modernity. Steel's hardness makes it the right stuff for tools that can spin, cut, and shave at high speeds. Steel is a vital organ of mass manufacturing. It is the stuff of car frames, bulldozer scoops, appliances, implements of every imaginable kind, and massive caches of weapons. Aside from concrete and stone, industry churns out more tons of steel than any other material. U.S. steel production in the 1990s has hovered around 90 million metric tons.

So central has steel become to so many products that

uniformity and reliability have been mantras for those in the steel industry. Failure because of flawed materials is simply too expensive. I-beams for a new high-rise in San Francisco or Tokyo must be able to hold the building up even in an earthquake. Steel used by automakers had better not cost them customers and embarassing recall by corroding, cracking, or degrading. The same goes for specialty items like the bearing steel used in the fuel turbo pumps of the space shuttles' engines, on which the lives of astronauts and billion-dollar satellites depend.

The imperative for reliability can make an industry conservative. Once a good alloy fits the bill for a specific application, talk of changing anything becomes heretical. That conservatism has worked well enough for erecting cities and furnishing the world with all manner of steel-dependent products. But in an era like this one, when each new generation of technologies makes more stringent demands on materials, the current roster of steel alloys already looks like old-timers' day at the ballpark.

The wish list for better steel alloys is a long one. The Navy wants stronger and tougher steels for the arrestor hooks without which their jets would be unable to stop as they land on aircraft carriers. The automobile industry can always use stronger steels to allow them to shave fuel-eating pounds from each car. NASA has called for better bearing steels for the fuel pumps of its rocket engines. So vast is steel's role in modern societies that any significant improvement in its properties can have revolutionary implications. If you make an alloy twice as strong as the best available today, you will need only half as much metal to build a city (provided the strong alloy is cheap enough!). A steel alloy that won't shatter in the chronic presence of reactive hydrogen atoms will furnish the key to building the future's fusion-energy reactors.

Ever since steel became a high-volume material for society, metallurgists have tried to sweat the last drop of technical performance out of it. Mostly, however, new steel alloys have been the happy and sporadic results of—so what's new?—trial and error. Materials researchers often refer to this method as the shake-and-bake or heat-and-beat approach to making materials. Yet there has never been a science-dominated approach to developing steel alloys that has met the imperious demands of industrial society.

That tried-and-true empiricism, however, has reached a point of diminishing returns. The chances of improving steel alloys significantly by such an approach are growing ever slimmer. It just takes too much time, too many trials, too many tests. Luck is the limiting factor. Yet materials researchers know that no one has tapped steel's ultimate capabilities.

By getting herself born in 1982, Greg Olson's daughter may have helped to set the history of steel, and of materials development in general, toward a new destiny. "When she was born, my time line suddenly expanded much farther into the future," recalls papa Olson, a research expert in steel metallurgy. At that time Olson was a materials scientist in his mid-thirties at the Massachusetts Institute of Technology, a citadel of materials research. Before that he had been a graduate student of Morris Cohen, one of the most revered elders of materials research.

Olson was trying to establish a track record in the field by uncovering more details about what makes steel alloys strong and tough. The tougher a steel alloy, the more of a beating it can take without deforming, indenting, or disintegrating. The stronger a steel alloy, the more weight it can support. So if you learn more about what the iron, carbon, chromium, and other elements in the steel alloy actually do, you might find keys to making better steels. Olson was in fact authoring the kind of papers that would help him move up the academic ladder from his modest post as a research associate. Becoming a father, however, changed his priorities: "I suddenly felt that I ought to be working on things that could make the world better for my daughter, not just on things that could earn me respect in my academic world."

What has come of his epiphany is far bigger than what he originally had in mind. His long-term goal has become nothing less than to reengineer the way humanity invents the materials of which society is built—not just steel, but all materials. Says Olson, "The reality of all of the materials that we have developed so far is that they have not been designed." Trial and error, serendipity, variations of traditional formulas, with an occasional dash of science—this is how most materials have been discovered and developed.

The empirical way of making steel is to prepare many alloy compositions like so many different cake batters. Each one of these is made by weighing out stipulated proportions of the alloying elements (80 percent iron, 12 percent chromium to make it stainless, and so on). The little clinking scraps, spheres, and other shapes with which these alloying elements start life are placed into a ceramic crucible and then heated until they melt into the searing molten alloy. The liquid metal looks and feels like a miniature sun. To get near it is to defy all your survival instincts, which are saying stay away. The melt, as metal people refer to the molten brew, is poured into a mold, where it cools and freezes into the solid alloy. The cooling may be a simple affair, or it might involve various tempering or reheating steps. In a tempering step, for example, the solid metal might be held for a while at a temperature of, say, 500°C. That way the metal remains loose enough inside so that certain internal atomic rearrangements linked to strength or ductility can occur within the alloy. A final cooling step then can lock those new atomic structures in place.

When you set out to make a steel alloy with a particular set of properties, you don't actually know what you have until you do some tests. You need to do some metallography so you can see what internal structure you actually have. You need to put a piece of it into a testing machine that will measure how much tensile force it can handle before breaking. You need to put it into another testing machine that can measure how hard it is. You might need to cool it down to frigid temperatures to see how brittle it gets. If you are systematic enough about your alloy formulations and property measurements, you may discern useful correlations between the metal's chemical composition, the way it is processed, and the physical properties of the resulting alloys. If you are really lucky, you will find a winner—an affordable, commercially attractive alloy.

For centuries discovering correlations like those have been the bread and butter of metallurgy and all materials development. That is how early metallurgists discovered bronze five thousand years ago. That is how Wallace Carothers discovered nylon in the 1930s. That is how two researchers at IBM's research facility in Switzerland discovered high-temperature superconductors in 1986—all of which proves that empiricism is not a bad way to go. Empirically discovered materials have been plenty good enough for building modern society.

But what was good enough rarely remains so for long. Steel is a case in point. To Olson contemporary steel alloys and most other materials are "highly inefficient." Their strength, durability, resistance to the elements, and most other measures of performance fall short of what steel alloys theoretically are capable of. To Olson the inefficiency of materials translates into an across-the-board drag on human progress. Bridges collapse, machine bearings wear, tires blow out, roads crack, metal corrodes. Engineers have always back-pedaled on their designs because of the limitations of the available materials.

Olson, perhaps the most avid champion—even philosopher—of materials design, sees this situation turning completely around. "Engineers no longer have to be limited by materials that exist," he claims. "They can design according to the materials they would like to have." Want to build a skyscraper 300 stories high? Get a materials researcher to invent a steel alloy strong enough to do the job. Want a semiconductor laser that emits light at short blue wavelengths so that you can pack twenty times more information onto a compact disc? Get a materials researcher like Federico Capasso to invent a semiconductor structure that can fill the order.

"If we could actually design materials, we ought to come up with materials that are a whole lot better than any we had before," Olson says. If scientists could design and make more capable materials, the things society makes and needs would go faster, last longer, pollute less, recycle more completely, store more information, transmit more data in less time, and generally do more with fewer shortcomings.

In this framework the feat of materials design that band-gap engineers have been pulling off for years is merely a single embodiment of what Olson has in mind. From quantum mechanical theory, band-gap engineers develop detailed blueprints of new semiconductor crystals. From the blueprints crystal growers grow the designed crystals. And quite often the crystals work just as predicted. This is scientific materials design.

Olson aims to prove this point with steel. Compared to steel, he says, "No material is in a better position to be designed from scientific knowledge, and no material is in more need of this approach." But there are good reasons why semiconductor researchers have succeeded in designing crystals well before steel researchers have demonstrated the same capability with metal. After all, even the quantum cascade laser (QCL), in all its five-hundred-layer glory, is a simple material compared to steel. The QCL is a member of the most idealized club of all solid materials: it is a crystal. Its complexity comes from the fact that its many layers have different chemical compositions. Unlike a diamond crystal, which is everywhere the same, the structure of a QCL crystal is like a stack of blankets of different thicknesses and colors.

A steel alloy is another animal. It is more like a compacted load of earth with sand particles, pebbles, larger stones, occasional blades of grass, roots, and God knows what else all jumbled together. The performance of each piece of steel depends not only on how its atoms are arranged into tiny crystal grains of different shapes and sizes, but also on how these accrete, bond, or jam together into ever larger structures that you could hang a concrete facade onto or make a high-speed cutting tool out of. It depends upon different structural rearrangements and transformations that can occur on the different-size scales. It depends upon what little films of minerals and other chemical structures end up on and between the grains. You cannot make steel with a molecular beam epitaxy machine.

Can steel be designed with the same finesse as semiconductor crystals? Olson has been betting yes for more than a decade. "Steel is probably unique in the extent to which we know the function of all its structural levels. We can have a list of quantitative structure-property relations [such as the relationship between grain size and strength]," he explains. "This means we have the tools to design steel alloys."

There are a few hurdles, of course. There are so many relevant structure-property relations to consider at one time that the complexity of the problem has made quitters of previous would-be steel designers. Each piece of metal comes from a metallurgical drama that plays out as an alloy's atomic ingredients are prepared, mixed, made molten, and then forced to react, coalesce, segregate, and otherwise interact as they take their final place in the finished alloy. So much physics and chemistry go into making steel that no one totally understands it. The chemical and physical forces at work invariably pull the action simultaneously in different directions, forcing the metallurgist to become a master of trade-offs. If you add one ingredient to make the alloy stronger, it becomes more prone to cracking than before. That's bad news if you are trying to make a better metal for the turbine blades inside a jet engine. Adding a different ingredient to reduce the alloy's brittleness may result in greater proneness to corrosion in the rain. Such an alloy is useless for anything that needs to serve on, in, or near the ocean. Finally, practical metallurgists know that the costs of metal-making and finished products must hew to commercial limits. You can make the world's most wonderful materials, but if no one can afford the stuff, they are as practical as the Hope Diamond.

Undeterred by these formidable obstacles, Olson was determined to forge a new approach, a new understanding. It would take a first-person leap of imagination: a steel designer would have to become steel the way *Star Trek*'s Mr. Spock almost becomes the person with whom he mind-melds. To hear Olson speak about steel to his expert colleagues is to witness a man who has mind-melded with metal. (See Figure 37.)

Olson's ambition to learn how to design steel is more than personal: he wants to begin a movement that will change the overall practice of materials research. What better way to begin a sustainable movement than to indoctrinate your students? That is why Olson's students train from day one to become materials designers in the only way possible. They can't learn this approach from books because such books haven't been written yet. So Olson's students, along with Olson himself, learn to become materials designers by being materials designers.

Olson began laying down the foundations for this philosophy of materials development in 1985 by founding the Steel Research Group (SRG) at M.I.T. with his mentor, Morris Cohen. Cohen, along with the likes of Morris Fine, was among the cadre of researchers who first defined the field of materials science and

---

Anatomy of Steel. Despite its uniform outward appearance, a piece of steel has a highly differentiated internal anatomy. The micrograph above is of a metallographic preparation of a piece of martensitic steel. Martensite refers to a particularly hard crystalline arrangement of iron and carbon atoms that can form when hot steel is quenched, or rapidly cooled. The martensite phase shows up here as the darker regions within a lighter-appearing matrix in which the steel has not converted into martensite.

GREG OLSON, ASM INTERNATIONAL

engineering. Members of the SRG included researchers from academic, government, and industry laboratories. The National Science Foundation provided the bulk of the initial funding. Ironically, the steel industry at the time was in such a trough that some of the SRG's original industry partners would go belly up the very next year. Then some of the smaller, more innovative steel companies such as Carpenter Steel came on board. Each year overall membership has shifted. But the SRG typically has involved several dozen researchers. This sounds like a big operation, but like most materials research it really

FIGURE 37

has been small science. Since its inception the SRG has spent about $10 million from government and industry sources.

The SRG seeks to develop a strategy to design steel alloys to order. If they could succeed with steel, they would try to apply the new design paradigm to other categories of materials. In time, Olson hoped, materials researchers would design all kinds of metals, polymers, and even concrete (actually one of the more complex materials on earth) with capabilities surpassing anything resulting from empirical discovery alone.

First things first. The initial technical goal of the SRG was to create steel alloys that are significantly stronger, tougher, and more corrosion-resistant than any of the empirically discovered alloys already out there: less steel would be needed in traditional applications like manufacturing cars; better steels also would increase the capability, reliability, and safety of spacecraft, aircraft, ocean craft, and other technologies that use expensive, state-of-the-art steel.[1]

Consider a single gain that NASA wanted so much that it began pouring money into the SRG soon after it formed. This saga began in earnest in 1986, when the price of material failure became graphically ingrained in contemporary culture. Over and over again on TV screens across the world, the space shuttle *Challenger* exploded seventy-two seconds into its ascent. The culprit was the elastomeric (rubberlike) O-rings between segments of an external fuel tank. During the especially cold launch on that fateful day, the O-ring had gotten too stiff and brittle to handle the tremendous stresses of takeoff. Hot combustion gases blew by the rings and ignited the fuel-loaded tank. What many people in the technical community in and out of NASA knew was that the rubber O-rings just happened to be one of many potential accidents waiting to happen.

Before the disaster, just as the SRG was getting off of the ground, a friend of Olson's at Rockwell International, a high-technology firm with a long history in developing aerospace materials and systems, suggested a perfect challenge for the would-be steel designer. There was another material-based time bomb in the shuttle fleet, a steel-based time bomb. In the heart of the main engines of the spacecraft are powerful fuel turbo pumps that inject frigid liquid oxygen into the engines, where the oxygen mixes with the fuel to enable combustion to occur. In those turbo pumps are bearings made of a steel alloy known as 440C. Engineers chose the alloy for its combination of strength, toughness, corrosion-resistance, and imperviousness to damage from the high levels of reactive hydrogen atoms in that environment.

It is a hostile environment at that. The bearings must support shafts that spin around thousands of times each minute at cryogenic temperatures below -300°F. All the while the assembly must pump extremely corrosive liquid oxygen into an adjacent engine that generates 30,000 horsepower in an area that could fit under an automobile's hood.

If these bearings fail, so does the pump. If the pump fails, the engines become a bomb. Even the 440C steel itself would become a fuel. Everything would ignite. NASA engineers were concerned about the potential. But they apparently had found ways of modifying the alloys so that the bearing would pass during ground tests of the engines. Safety entailed frequent inspections and replacements of the parts.[2]

That is why it was not a happy day when NASA engineers discovered that a shuttle had returned with a cracked steel component in the engine. Still, with no replacement material available to solve the problem, the only solution was to inspect and replace the bearings frequently. NASA materials engineer Steve Gentz told a *Chicago Tribune* reporter in 1989 that these inspections had become a significant source of cost as well as delay between shuttle missions.[3] A new and more reliable bearing steel would make a lot of people sleep better. To Olson it was a perfect target. And with new funding from NASA, it became the SRG's largest project.

Meanwhile, the SRG was running headlong into some

daunting obstacles of the attitudinal, sociological, and cultural sort. Designing materials involves such a breadth of techniques, theories, practices, and mind-sets that it bridges the yawning gulf between science and engineering. Olson and his colleagues even have a cute neologism for the pursuit: scieneering. To design materials, therefore, meant that a lot of people had to get together to work as a team. Moreover, producing a new breed of scieneers entailed pushing a new kind of engineering curriculum.

Inspiring teamwork and forging new curricula have never been small feats at universities. True, materials research already was born on a bridge between science and engineering, and M.I.T., Olson's academic home when the SRG first formed, was in principle receptive to the linkage. Still, Olson recalls encountering stiff opposition. For one thing, consider the likelihood of convincing a bunch of upwardly mobile researchers at a high-profile place like M.I.T. to become mere fish in a larger pond such as the SRG. Besides, Olson recalls, steel was "kind of 'out' as a research topic in universities." Most materials researchers in the academies had their sights on newer, more exotic materials such as light-emitting semiconductors and electrically conductive polymers.

Northwestern University was an exception. At the time it hosted the industry-sponsored Steel Resource Center, where steel companies and researchers could go in search of answers to broad technology and management problems. So here was a university with a tradition of interdisciplinary teamwork in materials research and a strong commitment to steel. It was perfect for what Olson had in mind.

Olson, too, seemed perfect for the visions of Stephen Carr, the head of the university's Materials Research Center (MRC). Carr already had been convincing himself for years that materials design would be the most exciting direction in materials research. Olson's steel lab was just the right addition to the MRC to put Northwestern at the forefront. Olson and Carr talked about their common visions. In 1988 the flirtation became a marriage: Olson transplanted himself and about twenty members of his M.I.T. lab to Northwestern's Materials Research Center.

With backing from his new academic home, Olson now had to confront an educational divide that transcended locality. Standard science and engineering curricula train students to analyze phenomena, to reduce things to their basic parts: people become organs; organs become cells; cells become molecules. Often lost in this kind of reductionism is a feel for the myriad ways in which the same parts can come together and produce unexpected results.

For example, how in the world does the property of stainlessness (resistance to rusting and corrosion) emerge from iron, carbon, and chromium atoms? Analytical minds are not fully equipped to excel in this more synthetic, holistic aspect of materials research. It takes more than analytical precision and mathematical acumen to get a feel for emergent properties that are more than a mere sum of parts. Worse than that, Olson believed that the traditional focus on analytical thinking literally had been disabling generations of students, stunting the kind of creative thinking and research the field cried out for.

Music composition, it seemed to Olson, was the model that could change the situation. When Olson was a graduate student at M.I.T., he supplemented his technical program and course work with music composition. Composing music is a procedure of synthesis, not analysis. To compose music is to put together different instruments, sounds, tones, silences, and rhythms into a whole that soothes, agitates, uplifts, saddens, pleases, or otherwise affects the listener. The composer must orchestrate the composition's progress from the beginning to the end of the piece. All the while the composition's properties—its flow, style, emotive effects—must merge into a singular performance.

To Olson musical composition holds more than just a metaphoric resemblance to designing materials. Musical composition and material design of the highest creative order partake of similar cognitive machinery. Both require a thorough knowledge of the parts involved and an intuitive grasp of how those materials come together into ever larger and more complex structures; both demand a sense of how the parts of a composition can merge into something greater than a sum of those parts. Olson's ideal materials designer, then, is an engineer with the training of a scientist and the soul of a musician.

This musical model of materials design shows up in what Stephen Carr affectionately refers to as Olson Diagrams (an allusion to the famed Feynman Diagrams that the late Richard Feynman developed to depict interactions among elementary particles). As Olson sees it, his diagrams are to materials designers what arrangements are to jazz composers and musicians. At the top of a typical Olson Diagram for steel are three headings with a handful of items underneath each: from left to right they are Processing, Structure, and Properties. Lines radiate from the specific items under each heading to specific items under the adjacent heading. These lines show how specific processing steps lead to specific material structures and how specific material structures relate to the material's physical properties. (See Figures 38 and 39.)

Take the Olson Diagram for nickel-cobalt steel, the kind of superalloy used to make the highest-strength aircraft landing gear. Under the Processing heading is a set of boxes that outline the way molten metal becomes a solid piece of an alloy: refine the raw molten steel; remove oxygen from the melt;

Olson Diagrams. The diagram for an ultrahigh-strength steel alloy shows how the technologically important properties of materials depend upon the complex internal structure that emerges as the metal undergoes processing. The generality of this conceptual framework for thinking about materials is made apparent by the second diagram, which depicts how the important properties of ice cream depend upon specific types of internal structures, which again depend upon the way ingredients are processed.

GREG OLSON

FIGURE 38

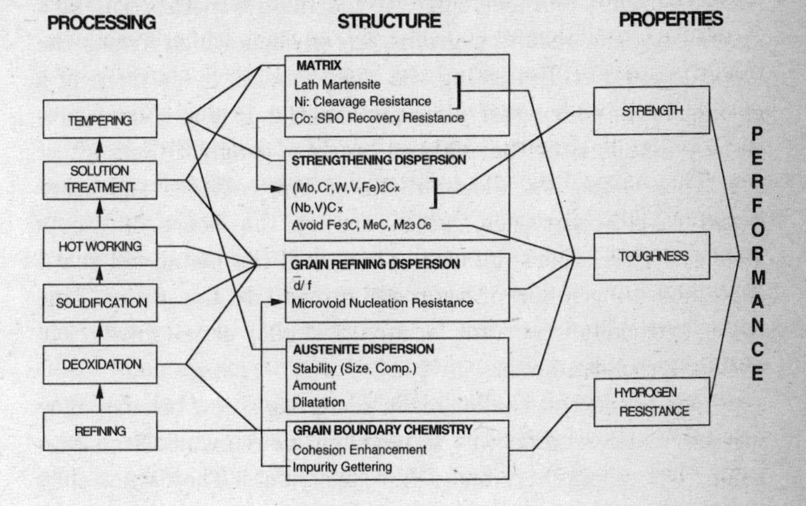

FIGURE 39

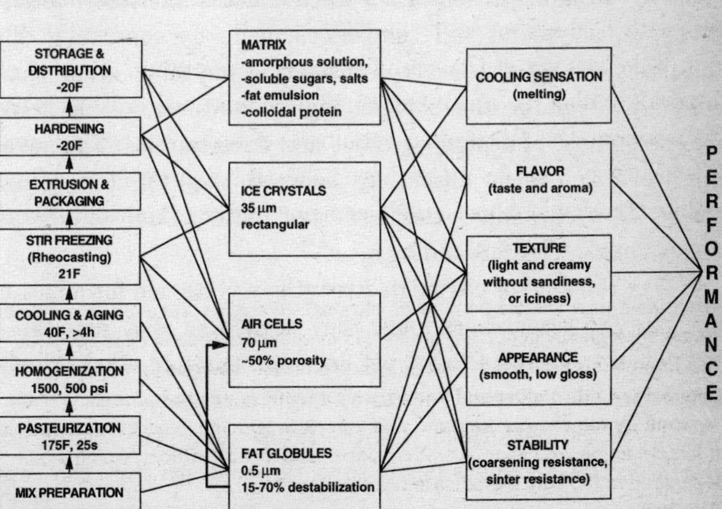

solidify; temper the solid metal. Each process step changes the internal structure of the metal. And each structural change increases or decreases the alloy's strength, toughness, corrosion-resistance, or formability. If the molten metal cools at a faster rate, the resulting grains are smaller, which means the metal is tougher. Reheating the cooled metal yields more of a strength-related crystal structure in the grains. Some processing details alter the cohesion between grains. Others affect the types, sizes, and distributions of iron carbides and other metal carbide particles, which are at the heart of steel's strength. Still others determine how well the metal will resist the embrittling influx of hydrogen atoms into the grains. This last is especially important for any steel alloy slated for duty in or near a rocket engine.

One of the first challenges in designing a new bearing alloy for the shuttle engine was to develop the relevant Olson Diagram. The researchers had a few constraints. They knew they would have to stick with a steel alloy akin to the high-chromium stainless-steel alloy already used for this shuttle application. The most general engineering approach would have been to consider all kinds of materials as candidates: nonsteel metals, ceramics, polymers, and composites made by combining different classes of materials. Considering anything other than stainless steel for the shuttle engine, however, would have entailed a host of design novelties and uncertainties. The novel approach the team ultimately adopted was the conceptual fusion of a chromium stainless steel with a high-toughness nickel-cobalt steel for landing gear.

For the kind of ultrahigh-strength steel needed for a better bearing, the researchers knew they needed an alloy noticeably tougher than the 440C alloy yet uncompromised in other critical properties such as resistance to wear and corrosion.

Once the performance of the future alloy was stipulated, the materials designers began to work backward from the venerable empirical route. From the performance they could infer which physical properties the steel must have. From the properties they could infer the kinds of internal architecture they had to realize. And from the architecture they could infer the details of chemical composition and processing that might actually create the structure.

Where there is inference, there also is a lot of theorizing and modeling. Where there is theorizing and modeling, there is a lot of calculating, too. That means computers. Materials design is in part a child of microprocessors capable of making quantitative sense of complicated theories. One of the most pivotal tools for the SRG goes by the name ThermoCalc™. It is a database and software system for calculating how an alloy's elements move, align, combine, interact, and avoid one another as their context changes from liquid metal through more viscous stages until they lock into place to form a solid piece of steel. When integrated with appropriate metallurgical models, it can calculate the likely size of grains, carbides, and other metallic and ceramic phases that form in alloys involving up to fourteen different atomic ingredients. It can calculate how different cooling rates affect the outcome.

"It was the first time we had a tool like that," Olson recalls thinking in 1985 when ThermoCalc™ first became available from the researchers in Stockholm who developed it. "It was one of the first things that made it possible to think about designing materials."

ThermoCalc™ itself depended heavily on decades of empirical observations and thermodynamic data gathered by thousands of researchers around the world. From the number-crunching with ThermoCalc™ and a battery of other computational tools emerge candidate alloy compositions along with a blueprint of the microarchitecture that the alloy ingredients must assume. The new bearing alloy candidate that ultimately emerged consisted of seven components: iron, carbon, nickel, cobalt, molybdenum, chromium, and vanadium, all interacting to maximize the alloy's strength, toughness, and hydrogen-resistance. In shorthand the alloy is written in weight percentage this way:

55.95Fe–22.6Co–12.0Cr–8.6-Ni–0.3Mo–0.25V–0.30C.

Now for a sampling of the details. As always, iron and carbon form the basic steel alloy. The SRG arrived at the specific amount of carbon, 0.3 percent by weight, by carefully examining experimental data that relates a steel's hardness to the distribution of carbide particles in the steel. Any more carbon and the steel would be too brittle. Any less would deprive the steel of the hardness necessary to handle the metal-on-metal wear inside the shuttle engine.

The 12-percent chromium component of the alloy makes the steel stainless by forming an invisible film of chromium oxide on the metal's surface, preventing oxygen atoms from finding iron atoms at the surface to form iron oxide, otherwise known as rust. Even when the surface oxide scratches away, there is enough chromium for a new oxide surface to form immediately. With less chromium, the stainless property would degrade; with more, strength and toughness would suffer.

Nickel and cobalt—primary ingredients in advanced aircraft landing-gear steels—are necessary to promote both strength and toughness. Precisely how much of these alloying ingredients to include was the question. For one thing, these elements control the temperature at which an important structural transformation occurs in the steel during cooling. But the researchers wanted to keep this temperature high enough to achieve a specific microstructure related to strength within the alloy's grains. Calculations with ThermoCalc™ yielded a small range of nickel and cobalt compositions that would be suitable. Only one of those compositions—22.6 percent cobalt and 8.6 percent nickel, it turns out—would also allow for a desirable toughening mechanism to occur when the solid metal is tempered at about 500°C.

Meanwhile, the new alloy had to have ultrahigh strength, more than would come from the iron, cobalt, nickel, and chromium carbides. From experience and previous experiments, the SRG knew that molybdenum and vanadium should be able to do the job by joining with carbon in specific arrangements to form complex chromium-molybdenum-vanadium carbides. The trick, however, was to find the compositions that would lead to the formation of many more small carbide particles rather than a lesser number of larger ones. The smaller particles lead to stronger alloys; the larger ones to more brittle alloys. The difficulty here is that different amounts of vanadium and molybdenum have a characteristic "driving force" for producing the small carbides at the expense of the larger ones. Here, the team relied on both experimental data and thermodynamic calculations to come up with the specific amounts of the two alloying agents: .3 percent molybdenum and .25 percent vanadium.

So there it finally was:

55.95Fe–22.6Co–12.0Cr–8.6-Ni–0.3Mo–0.25V–0.30C.

It was perhaps the first formula of a designed steel alloy. But was it in fact any better than the alloy it was designed to beat? This is where SRG corporate members like Carpenter Steel come in. Their metallurgists can fill the metallic prescriptions that Olson and his students come up with. Into a small test crucible went the right proportions of alloying elements to make the smallest sample they make—17 pounds. When the SRG team tested samples from the resulting 17-pound ingot, they were heartened by the results. The alloy was indeed twice as tough as the bearing steel NASA had been using. They had, in fact, designed a steel alloy with at least some of the right stuff for the bearing application.

It was a moment of private triumph, but the result had

more potential than immediate value; the prototype material could not be processed in ways suitable for many applications. Because of a lower than expected structural transformation temperature, the alloy required a processing step at very low temperatures to achieve the desired internal microstructure. Although bearings still could be fabricated this way, the processing of alloy compositions would be more difficult and costly than that of current bearing steels. That partly explains why the particular seven-component alloy that Olson, his students, and his SRG colleagues had designed and that Carpenter Steel had made never made it into a space shuttle pump.

It didn't make it into the pumps for another reason. Before the SRG could try to design a more processable composition, NASA funding for the project disappeared suddenly as a result of the agency's wrenching post-*Challenger* reorganization. Still, in 1995, NASA did end up patenting the alloy that the SRG designed. NASA noted in a technical brief on the alloy that it represents a revolutionary new breed of scientifically designed materials.

In the late 1980s Olson began choosing more generic design targets. Rather than trying to design an alloy whose application would be limited to specialty applications, like fuel pumps for spacecraft, Olson decided to aim at alloys that would be more widely used. It was getting back to the basics. The huge technological and economic potential of such an advance is self-evident in this country, which produces almost 100 million tons of steel each year.

One project emerged from a continuing collaboration Olson has with the Newman-Haas Indy car racing team—an outgrowth of his involvement in amateur sports-car racing. The chief engineer of the team, Brian Lyles, wanted to reduce the weight of the steel gears in the cars' drive trains. Gear steels are harder on the outside so that their teeth won't wear down and less hard inside so that the gears won't crack from fatigue. The gear steels everybody was using had been around since World War II. Some modest improvement in the performance of these alloys had been achieved by reducing the amount of contaminants in the alloys. But there had been no dramatic improvements.

In 1991 Olson adopted the goal of improved gear steels as a project for his materials design class. Racing cars are not the only place where improved gears would do some good. For every pound shaved off of a gear that goes into a commercial airliner, for example, the economic payoff would be about eighty dollars. There are about two tons of gears on a Boeing 747. Instead of flying an extra ton of gear steel all over the place, you could take an extra ten paying passengers. Similar logic suggested that military aircraft made with the steel could carry higher payloads or fly farther.

The kernel of the challenge was to make a steel that is harder in the surface (case-hardened), where the teeth of the gear must withstand a lot of metal-on-metal contact, and more pliant internally so that the gear will be able to take the external stresses and internal strains that come with their use. By hitting a hardness level known as Rockwell 70 (named after the test protocol that has become the standard), the team believed it also would increase the alloy's resistance to fatigue. If they could realize these combined improvements, they stood a good chance of cutting the gear weight in half. Creating the hardness gradient entails creating a corresponding gradient in the amount of carbon in the alloy. There should be less carbon inside and more on the outside, by the gear's teeth.

With an Olson Diagram, a century's worth of data, physical modeling and theorizing, and patient hours in front of computers, the design class came up with a design. The strategy was to increase the hardness of both the case and the core, but without also making the core more brittle and susceptible to corrosion under stressful conditions. As in existing gear alloys, the new alloys would require a "carburizing" step to inject extra carbon into the surface of the metal. Carpenter did a 17-pound melt. Again, the result was close to the target. The surface of the alloy was just shy of the hardness level needed, while the inside was a bit too hard.

Still, the exercise confirmed Olson's faith that designing steel with stipulated properties was becoming possible. In 1996 the SRG sent off a second prototype with improved properties to the Newman-Haas racing team for testing. As often happens in materials development, the work may prove useful in unintended ways. After the student project won a national materials design competition, the properties of the alloy caught the attention of a Navy research program manager who hopes the SRG will be able to design stronger and tougher steels for the components in the landing gear of their aircraft.

The most futuristic project was inspired by the film *Terminator 2: Judgment Day*, in which a future fascistic regime sends a robotic killer back in time to murder a boy who will grow up to become a military leader for the opposition. The robot is made of a biomimetic, self-healing metallic alloy that can return to its former shape even after being blown to bits. After seeing the movie and considering its potential to inspire students, Olson asked himself a question: "Could we actually make a self-healing material?" The challenge went over well in his design class.

One of the first leads for a design came from the same abalone shells that had inspired Mehmet Sarikaya of the University of Washington and others of the biomimetic persuasion. The shell's brittle ceramic is reinforced by a relatively tiny amount of a rubbery protein and carbohydrate-based polymer. If a crack begins to run through the ceramic, the rubbery reinforcement bridges the crack and helps resist further separation.

Olson wondered whether a similar strategy could work for a class of metals known as high-temperature intermetallic superalloys. Even though these metals work exceptionally well at the high temperatures found in places like jet engines, they have a tendency to be brittle at lower temperatures. Engineers had told him that they would like to find a way of reducing that vulnerability. How could the abalonelike toughening mechanism be realized in a high-temperature intermetallic alloy?

One possible key resides in a remarkable class of metals—called shape-memory metals—that can be severely bent out of shape but then returned to their original form by applying heat. It is a favorite class of materials of the smart materials tribe. These metals have been used to flip up headlights on sports cars and as mechanical elements in torpedoes. In a metallic setting a shape-memory metal could behave like a high-strength rubber cement analogous to the protein-carbohydrate glue of the abalone shell.

Let's say that at low temperatures overloading the metal causes microscopic cracks to form in an intermetallic alloy containing a reinforcing distribution of shape-memory metal. At the low temperature the shape-memory component can stretch to bridge the cracks just like rubbery reinforcement in the seashell. But now, with the alloy heated up, the shape-memory components would reverse their deformation, causing the cracks not merely to get no worse but to actually close. Moreover, by designing the shape-memory alloy so that it actually creeps less under stress than the surrounding superalloy, the reinforcement could provide enough clamping to allow rewelding of the cracks at high temperature, thus completely healing the internal cracks.

One of the most basic requirements for the design is to identify a superalloy composition and a shape-memory alloy composition that don't eat each other up. The challenge is sort of like putting hot milk and chocolate syrup together and hoping that the syrupy quality of chocolate doesn't completely give way to a pure fluid. The initial vision is to start with titanium aluminide intermetallic alloys whose combination of strength and low density (lightness) have made them important for military aerospace applications. To achieve the self-healing feature, the SRG proposes to reinforce the matrix metal with a titanium-nickel shape-memory alloy. The hard part is finding a way of getting the right chemical compatibility of the superalloy and the shape-memory phases to avoid combining ingredients completely into chocolate milk.

The research required in this project runs the gamut. It includes fundamental quantum mechanical calculations for determining the factors favoring stability of the shape-memory alloy. It includes higher-level mechanical calculations to reveal the various interactions that will affect the toughening behavior of the shape-memory alloy when cracks form in the titanium aluminide matrix. It also includes some experimental determinations of how alloying agents such as palladium, hafnium, and zirconium might yield an alloy in which the self-healing behavior will be present even at higher temperatures. That would be helpful for the hot venues such as jet engines in which these alloys might be used.

Olson is the first to admit that any aerospace engineer who depended on an engine part made of an alloy designed in this clean, *in-silico* manner without testing the prototypes is the kind of person willing to bungee-jump without testing the cord. There will never be substitutes for making prototypes of the designed alloys, analyzing their structure, measuring their properties, and monitoring their performance in their intended applications. That is why Pratt & Whitney, one of the world's major jet-engine manufacturers, is part of the project. The company will make the prototypes.

Since Olson's team is sending *designed* alloys to Pratt & Whitney, the expensive and time-consuming process of testing prototypes should be considerably shorter than usual. Rather than working up the twenty or so prototypes that come out of a standard empirical approach to developing a new alloy, the materials design approach ought to conclude after three prototypes. This kind of streamlining, Olson hopes, will reopen imaginations of materials researchers that had long been closed by fears of development costs.

Metal alloys are not the only kind of material that Olson,

Carr, and others have their sights on for this kind of R&D strategy. Although they see materials design as applicable to nearly all materials, it might well be best suited to materials such as steel that already have a long research history and thus an advanced knowledge base on which to build. But at Northwestern there are Olson Diagrams for advanced cements and concretes called "polymer-reinforced hydrate ceramic composites"; "case-hardenable rubber-toughened polymers" that could yield sophisticated polymer structures capable of serving as gears (for such high-volume items as windshield-washer motors and automatic car-window motors) and other products traditionally monopolized by metals.

Olson even envisions "superalloy ice cream," a material with a potential annual market of $9 billion. The Olson Diagram for ice cream looks like all of the others, except that some of the headings are more familiar—mix preparation, homogenization, flavor, cooling sensation, texture, etc. Just as in any other material, the properties of ice cream are determined by its internal structure. For ice cream that translates into details such as the size and distribution of ice crystals, fat globules, and air cells (voids) and the way sugars, salts, fat, and protein have dissolved and distributed in the milk-cream matrix. From the perspective of metallurgical and materials science, ice cream is just like a superalloy. Because of dietary and public health trends, foodmakers have been pressured to remove traditional food constituents such as fats whose material properties partly have determined the evolution of processing equipment and protocols. Mass producing acceptable fat-free food takes the skills of materials scientists. It stands to reason that materials researchers will probably have an increasingly important role in food industries.

"This is a glimpse of the future of materials design," Olson says. He says this of the kind of steel alloy design that the SRG group has been pioneering as well as of Capasso's MBE-grown crystals. Steel is one of the most brutish materials in use; band-gap engineered crystals are among the most delicate. Together, they stand like bookends for the entire materials spectrum. So it is reasonable to think that materials researchers will eventually apply their talents to all the types of materials in between.

The generic ability to design materials to order would be a world-historical technological advance. The ability to design better materials is the culmination of centuries of empiricism, analysis, synthesis, and theorizing about materials. Better materials mean better technology.

Better materials can also provide the fixes for many of the technological messes that have emerged from previous, less informed generations. But if the current generation of research and development succumbs to a similar short-sightedness, new materials could harbor dangers more serious than any we face today. Lead-containing salts made paint nice and white, but they ultimately became a huge public health problem. Chlorofluorcarbons (CFC), now pegged as the murderous chemicals that have been destroying stratospheric ozone molecules, were deemed benign wonder chemicals when they first became available in the 1930s. The good news is that materials researchers have learned from these experiences, and they are far better prepared to avoid repeating these mistakes. To be able to design materials from atomic scratch is to be able to dream up new stuff and to make those dreams come true. It deserves to stand among the few major turning points in the history of human control over the material world. To stone flaking, harnessing fire, the rise of science, and the emergence of materials science and engineering, add materials design as the latest and perhaps ultimate turning point.

Although the ability to design materials is new, Olson knows that it already has had some effect on his daughter. Taped on a window in his office at Northwestern University is one of her first-grade quiz papers. One of the questions is "Name two machines." Her answers: car and atom probe. An atom probe is a powerful analytical device that her father has used to reveal a material's two-dimensional or three-dimensional anatomy with atomic resolution. Now the task is to push the materials design paradigm along so that the periodic table will yield more of its latent potential for addressing real societal problems and ambitions. Then a father will have fulfilled his promise to make the world a better place for his daughter.

EIGHT

the STUFF THAT DREAMS *are* MADE *of*

When framed inside a book chapter, Greg Olson and Federico Capasso might stand out as rare virtuosi among a population of colorless journeymen. In fact, they are standouts only for the moment. They are like scouts of more dexterous next-link hominids amidst the first stone-flaking progenitors. They are representatives of what is to come.

But those two, because of their respective specialties—solid-state semiconductors and steel—bracket the entire materials spectrum. Capasso's multilayered crystal structures—grown atomic layer by atomic layer via molecular beam epitaxy and weighing all but a tiny fraction of a gram—are among the finest constructions possible in the materials spectrum. Olson's steel alloys, though designed in part with atomic detail, represent the other end of the spectrum, where the sophisticated, predesigned microstructures of these alloys are realized by clever control of the microstructures in multipound or multiton batches of alloys.

Moreover, the know-how, methods of synthesis and

processing, theories, and computational and analytical tools on which the successes of these materials designers depend are improving at an ever accelerating rate. It is inevitable that the materials design approach that Capasso and Olson have vindicated in their work will sweep the rest of the materials spectrum, including polymers, ceramics, composites, and ice cream.

If you talk to enough materials researchers, read enough of their papers, hear enough of their presentations, and visit enough of their laboratories, a surprisingly coherent message emerges from what otherwise appears to be a community as crowded and cacophonous as a kitchen's miscellany drawer: the materials research community has reached a new plateau of unprecedented power in its ability to understand, control, and manipulate the material world.

Do you need a lightweight alloy that remains strong at the searing temperatures of a scramjet operating at many times the speed of sound and able to withstand the frictional offensives of atmospheric molecules scouring the material like supersonic sandblasting? Do you need a polymeric material compatible with brain tissue that in the presence of specified levels of a specific neurotransmitter will release pharmaceutical agents previously loaded into the polymer? How about concrete ten or twenty times stronger than any available now? Call in your local materials designer.

Mark Eberhart, a materials researcher at the Colorado School of Mines, where most work focuses on ways of harvesting the rawest of ingredients from the world, characterizes the emergence of materials design this way: "I consider what is going on in materials by design as comparable in scope and importance to the beginnings of the scientific method. Now we are at a point where we can do computer simulations and use these as hypotheses in materials development."[1] This virtual experimentation saves the time and money once spent on dabbling with physical prototypes. This is akin to the evolutionary leap in which brains simulated the consequences of potential actions before hazarding the action itself: thinking ahead.

The materials design paradigm is permeating the materials science community, Eberhart observes. "It is simply the way in which we are training students." With computers everywhere getting more powerful at doing the complex calculations required by materials science, the new ethic seems destined to proliferate. Materials designers equipped with these powerful computers are learning to integrate the mountains of previous theoretical and experimental work with their experience-honed intuition.

Eberhart got his own start in materials research when he was building kayaks in the 1970s with DuPont's then new superpolymer Kevlar, which had a tendency to peel away from the kayak's inner foam forms. "When I called the DuPont guys and asked why the stuff was failing, they didn't know," Eberhart recalls. He decided that materials researchers ought to be able to answer questions like that and that he would become such a person.

Now, fully smitten by the simulation and materials design bugs, Eberhart takes on metallurgy problems that only computational approaches have much hope of tackling. For him, one particularly challenging problem is designing titanium aluminide intermetallic alloys (the same type of metal that the SRG is considering for self-healing metals) in the on-again, off-again (now very off) National Aerospace Plane (NASP). Designers of the NASP intended their aerospace craft to get into and back from orbit as easily and readily as a plane gets into and out of the jet stream. That would require materials capable of withstanding the heat and friction of multiple transatmospheric passages. Yet these same materials also would have to be as lightweight as possible since every extra pound of the vehicle represents an enormous cost in terms of money and/or reduced payload. The problem with titanium aluminides is their tendency to become more brittle as they absorb hydrogen from their surroundings (this was part of the problem Olson had taken on with the space shuttle's bearing steels). Hydrogen atoms would be abundantly available during each and every NASP trip.

Even though Congress killed the NASP program in 1994, the very fact that researchers like Eberhart thought they could build such a vehicle is testimony to their faith in materials design. It would fly only if materials designers systematically invented the right stuff.

The emergence of materials design will be felt not only by the materials designer and maker, but it will also have social and geopolitical consequences. It will become part of the equations governing the military and economic balance of power, local and global environmental health, resource management, and the quality of life for an expanding population. Those countries with the more talented materials designers will have the edge.

The drivers of innovation in materials will remain largely the same. Hypersonic aircraft such as the NASP is only one of countless examples of how engineers look to new materials to enable new technology. The need for new materials runs through anyone's list of high-technology desiderata. Faster computation depends in part on new semiconductors in which electrons move faster. Faster electronic materials mean faster computers, which mean more reliable weather forecasting and climate modeling, quicker design of more fuel-efficient vehicles, and more fruitful discovery of drugs by molecular modeling. New kinds of optical fibers, some of which have rare earth atoms such as erbium that have the curious ability to amplify light passing through them, mean more reliable, less complicated long-distance communication systems. New alloys, composites, and ceramic materials can mean a new system of transportation that moves people faster, more comfortably, with fewer accidents, with less energy and pollution. Armor that can withstand ever more powerful armor-piercing munitions has everything to do with new materials. Medical implants, prostheses, and artificial organs are also on the research agendas of teams of materials scientists specializing in biology and medicine. Name the technology and materials science is there in the forefront.

Pushing old technologies and developing new ones is just one of the drivers for new materials. Other factors are the failure of current materials to live up to expectations; the relentless market pressure to offer new and improved products; the military need for ever more capable materials and technologies; the academic honor of creating a material with superlative properties; and the societal benefit of having less expensive, more energy-efficient, environmentally safer materials.

But materials science isn't simply at the mercy of the prevailing social winds. The sophistication of researchers like Capasso and Olson allows them to anticipate and steer some of these driving factors. Engineers no longer need to design with the limitations of materials in mind; they can design technologies with materials they would like to have.

This is a kind of engineering bravado that used to be relegated to the minds of science fiction writers, who could simply imagine fantastic materials and technologies. Of course, the achievements of contemporary technology often dwarf the wildest imaginings of a century ago. Now the distance between thought experiment and real experiment is often measured in days or weeks rather than in decades.

On the other side of the Janus face of technology is the potential to design new materials that the world may eventually come to regret. In the 1930s Thomas Midgely, a chemist then working for Delco, developed tetraethyl lead. The primary concern was to come up with an antiknock agent to prevent military planes from stalling out, but it became a staple for the gasoline industry since it also improved the performance of automobile engines. Midgely also invented chlorofluorocarbons

(CFCs) as a refrigerant that would be safer than the ammonia and sulfur dioxide then in use. Only decades later, with production and distribution of these chemicals spanning the globe, did their darker side become apparent. Specters of lead poisoning and the destruction of stratospheric ozone molecules by CFC molecules have been among the gravest environmental and public health concerns in the late twentieth century.

There are plenty of others. The list of materials whose troublesome properties have come to outweigh their useful ones is quite long. Traditional lead-tin solders, though convenient to use since their melting point is low, are on their way to becoming completely phased out and already are taboo for most plumbing applications, where their past use has contributed to widespread exposure to toxic levels of lead. Asbestos is a fine thermal insulation material but also a dangerous carcinogen if its fibers are inhaled. Similar dangers may lurk in the reinforcing fibers and flakes made of other synthetic ceramic and mineral materials whose production could increase if composites become as commonplace as many engineers would like. Likewise, medical implant materials have improved the quality of life for millions of people, but some of their negative side effects are becoming apparent. In response to an article in *Science News* about attempts to make synthetic diamond coatings, one letter writer feared that an age of synthetic diamond, and the superhard and degradation-resistant coatings that could come of it, might produce a persistent waste problem. After all diamonds are forever.

Nevertheless, there are reasons to be optimistic. Materials developed before the environmental movement were driven by convenience, bottom lines, and ready availability of ingredients. Environmental issues were seldom seen on the checklists of materials researchers. That no longer is true. Designing for the environment is becoming integral to the mind-set of influential members of the materials community, the regulatory community, and users of the materials. In Germany, for example, automakers are responsible for recycling all of the materials that go into their cars.

One promoter of this environmental perspective is Robert Laudise, adjunct chemical director of Lucent Technologies. In the late 1950s he helped develop synthetic quartz for radio and electronics technologies, and he rose through the ranks to become head of Bell Lab's materials R&D. One way Laudise has tried to change the materials research culture is by requiring those under his wing to build environmental factors into their work at the earliest research stages so that potential problems will show up sooner rather than later. And one of the first things Laudise did in 1993, when he took over the helm of the *Journal of Materials Research* (one of the field's most visible publications), was to organize a special issue devoted to "Materials and the Environment."

Laudise is not alone in his desire to foment a new ethic in the materials community. Many companies such as 3M and Alcoa now nip the development of new materials, products, and technologies in the bud if there is potential for environmental harm even in the most promising products or materials. Companies that are realistic about regulatory action, public perception, and the growing cost of dirty operation and waste disposal are actively curtailing their pollution production and use of hazardous chemicals and materials. That means they have researchers, including materials researchers, who are developing a "cradle-to-grave" kind of thinking. This purview extends from raw materials extracted from the earth to the processing of these materials into products to the service lifetime of those materials and products to the endgame of recycling and disposal.

Even the enormous military infrastructure, which has often recklessly ignored the environmental consequences of some of the most hazardous materials ever created, is taking on a greenish tinge. The raison d'etre of the Hanford Nuclear Reservation in Washington State has shifted from making the nation's military plutonium to finding chemical and materials-based solutions to clean up the mess left in the wake of Cold War weapons production.

This sea change of thought and practice cannot ensure that there are no potential environmental disasters unwittingly being brewed in contemporary labs. But now materials researchers, government planners, and corporate managers are addressing the environmental and public health issues they had ignored or punted in the past.

Environmental risks notwithstanding, thinking about the technological future can still induce the dropped jaw and the starry eyes. Every week reports in technical journals like *Applied Physics Letters*, the *Journal of the American Chemical Society*, *Luminescent Materials*, and *Tissue Engineering* nudge and push the materials envelope just a little more. The unpublished results from company and government laboratories could be going even further.

What's more, progress is accelerating in each part of the materials research triumvirate: synthesis, analysis, and theory. The result is a kind of bootstrapping. Computational tools help steer synthesis and make sense of analysis. The ever better ability to synthesize structures to theoretical order and the data from ever more sophisticated analyses help refine the theories and simulations running inside the computers. Perhaps the most self-reflective example of this process is the use of silicon-based computer chips to design better and more powerful silicon chips, which, in turn, will be used to design the next generation of chips, perhaps ones made out of a material like gallium arsenide or even synthetic diamond.

It is possible, of course, that all this wondrous wizardry

with materials will merely color the history of technology but not radically reshape it. But given the startling technological upheavals of the past fifty years alone, contemporary materials science is likely to have as profound an effect on posterity as did that original act of materials engineering in eastern Africa's Rift Valley, where the sound of stone against stone first snapped into the Paleolithic air.

REFERENCES AND NOTES

This book emerged over a period of about five years, during which time I covered the field of materials science and engineering as a science writer. The book's themes come from a synthesis of thousands of individual components: visits to laboratories, interviews, technical papers, oral presentations at meetings, demonstrations, press conferences, and plenty of bookish research. To even attempt to list half of the sources that played some role in this book would, I think, fatten the spine without adding real value. In what follows, I offer a bibliographic essay of sorts that lists some general sources pertinent to each chapter and/or footnotes referring to the more specialized sources for particular points or claims.

Introduction: Stuff, Stuff, Everywhere Stuff

1. Cyril Stanley Smith, *A Search for Structure* (Cambridge, Mass: MIT Press, 1981).

Chapter One: Earth, Air, Fire, and Water

A good starting point for contemporary thinking about the earliest hominid relationship to materials is *Making Silent Stones Speak: Human Evolution and the Dawn of Technology* (New York: Simon & Schuster, 1993) by Kathy D. Schick and Nicholas Toth. Toth also has written a series of articles in *Scientific American* on early technology and the use of materials. A fascinating and informative source on the use of materials in earlier times is *Smithsonian's Timelines of the Ancient World: A Visual Chronology from the Origins of Life to 1500* (New York: Dorling Kindersley, 1993), edited by Chris Scarre. This oversize book is exquisitely and expensively brimming with pictures, diagrams, and graphs prepared by a platoon of authoritative scholars who have made it usable as a guided tour through the historical use of materials complete with a pictorial museum's worth of artifacts.

There is a large literature on the early human use of pottery, metal, and glass—the products of earthy ingredients treated with fire. The next stop, after technological encyclopedias and other more general standards like Jacob Bronowski's *The Ascent of Man* (Boston: Little, Brown and Company, 1973), is more and more specialized parts of the literature. A general consensus about the outline of this history has been available for decades, which means that most contributions today tend to shift or add details. The writings of Cyril Stanley Smith provide authoritative details, especially regarding metallurgy, put into beautiful, sweeping historical and philosophical contexts. See his *History of Metallography: The Development of Ideas on the Structure of Metals before 1890* (Cambridge, Mass.: MIT Press, 1988), which was one of the inspirations for this book. *A Search for Structure* (Cambridge, Mass.: MIT Press, 1981), is a wonderful collection of Smith's essays on science, art, and history that provides an impressionistic view of his philosophy of materials. Also good on metallurgy is Theodore A. Wertime. See, for example, his "Man's

First Encounters with Metallurgy" in *Science*, 4 December 1964, 1257–1267. Also valuable to me was an unpublished source entitled *Notes on the History of Metals and Ceramics*, which Professor Morris Fine of Northwestern University compiled for his courses in the early 1970s.

A Short History of Glass (New York: Harry N. Abrams, Inc., in association with the Corning Museum of Glass, 1980) is a readable, authoritative, and informative entry into the history of glass written by the curator of the Corning Museum of Glass, Chloe Zerwick. For early ceramic and fired-clay technology, see anything with the names of W. David Kingery or Pamela Vandiver, which are usually tough reads but worth the effort. An example is Kingery's and Vandiver's *Ceramic Masterpieces: Art, Structure, and Technology* (New York: The Free Press, 1986). Also see collections of research writeups such as *Early Pyrotechnology: The Evolution of the First Fire-Using Industries* (Washington, D.C.: Smithsonian Institution Press, 1982), edited by Theodore and Steven Wertime, which is a proceedings of a seminar on early pyrotechnology.

1. The following fictional vignette depicting the birth of Paleolithic stone tool technology is derived from the modern corpus of archaeological and paleoanthropological data and theories. Paleoanthropologists themselves resort to this kind of communication. See Schick and Toth, *Making Silent Stones Speak*, 77–78, 147–149, 187–189, and 225–227. Just which genus of early hominid, *Australopithecus* or *Homo habilis*, was the first to use tools remains a point of contention. For a recent articulation of the debate, see Randall L. Susman, "Fossil Evidence for Early Hominid Tool Use," *Science*, 9 September 1994, 1570–1573.

2. Cyril Stanley Smith. "Metallurgy as a Human Experience," in *Metallurgical Transactions A*, April 1975, 606–607.

3. Ibid.

4. Paul Rado, *An Introduction to the Technology of*

Pottery, 2d ed. (New York: Pergamon Press, 1988), 2. Some historians credit hunter-gatherers in Japan about twelve thousand years ago, not agriculturists in the Fertile Crescent, as the ones to first make pottery vessels. See, Scarre, *The Smithsonian Timelines*, 65.

5. Betty Jo Teeter Dobbs, *The Foundation of Newton's Alchemy or "The Hunting of the Greene Lyon"* (New York: Cambridge University Press, 1975).

6. Theodore A. Wertime, "Man's First Encounters with Metallurgy," *Science*, 4 December 1964, 1257–1267.

7. See, for example, Cyril Smith's "Materials and the Development of Civilization and Science," *Science*, 15 May 1965, 911. The quote comes from Smith's essay "Matter versus Materials: A Historical View," which was reprinted in *A Search for Structure*.

8. T. A. Rickard, *Man and Metals*, vol. 1 (New York: McGraw-Hill, 1932), 127.

9. Leslie Aitchison, *A History of Metals* (New York: Interscience Publishers, 1960).

10. See, for example, Bronowski, *The Ascent of Man*, 128.

11. Nicholas Reeves, *The Complete Tutankhamun* (New York, Thames and Hudson, 1990).

12. Cyril Stanley Smith, *Science*, 14 May 1965, 913.

13. ———. *A Search for Structure*, 1–67.

14. Zerwick, *A Short History of Glass*, 8.

15. Ibid., 12. See also Pliny the Elder, *Natural History: A Selection* (New York: Penguin Books, 1991), 361–363.

16. Zerwick, *A Short History of Glass*, p. 25.

17. Lewis Mumford, *Technics and Civilization* (New York: Harcourt, Brace & World, 1963), 124.

18. Ibid.

19. For history of early plaster, see M. James Blackman, "The Manufacture and Use of Burned Lime Plaster at Proto-Elamite Anshan (Iran)" in *Early Pyrotechnology: The Evolution of the First Fire-Using Industries*, eds. Theodore and

Steven Wertime (Washington, D.C.: Smithsonian Institution Press 1982), 107–115.

20. Ibid., 105–135.

Chapter Two: New Stuff in the Old World

A good starting point for the history of nineteenth century technology is the enormous multivolume classic aptly entitled *A History of Technology* (Oxford University Press: New York, 1954–1984), edited by Charles Singer, E. J. Holmyard, A. R. Hall, and Trevor I. Williams. Volumes IV and V focus on the nineteenth century. On the early days of polymers, see especially Robert Friedel's *Pioneer Plastic: The Making and Selling of Celluloid* (Madison: University of Wisconsin Press, 1983) and the same author's excellent booklet-length social interpretation of an exhibit on material at the Museum of American History: *A Material World* (Washington, D.C.: The National Museum of American History, Smithsonian Institution, 1988). Stephen Fenichel also has compiled a fascinating social history of plastics in *Plastic: The Making of a Synthetic Century* (New York: HarperBusiness, 1996).

Fred Aftalion's *A History of the International Chemical Industry* (Philadelphia: University of Pennsylvania Press, 1991) lays out the emergence and spread of a chemical industry from its nineteenth-century roots. Several chapters on metals in James Edward Gordon's *The Science of Structures and Materials* (New York: Scientific American Books, 1988) have excellent accounts of nineteenth- and early twentieth-century metallurgy.

For the history of chemistry, see William Brock's *Chemistry* (New York: W. W. Norton & Company, 1992). It is an excellent springboard into the literature, as is J. R. Partington's 1937 classic *A Short History of Chemistry* (New York: Dover Publications, 1989). Same for Hugh Salzberg's *From Caveman to Chemist* (Washington, D.C.: American Chemical Society, 1991) and David Knight's quite readable *Atoms and Elements* (London: Hutchinson & Co., 1967).

1. Chloe Zerwick, *A Short History of Glass* (Harry N. Abrams, Inc., in association with the Corning Museum of Glass: New York, 1980), 50.

2. David Kingery, "Ceramic Materials Science in Society," in *Annual Review of Materials Science* (Annual Reviews Inc., 1989), 5–8.

3. David Kingery, "Looking to the Future," in *Ceramic and Civilization: From Ancient Technology to Modern Science*, ed. David W. Kingery (Columbus: The American Ceramic Society, 1985), 378.

4. David Kingery and Pamela B. Vandiver, *Ceramic Masterpieces* (New York: The Free Press, 1986), 195–206.

5. Cyril Stanley Smith, *A History of Metallography: The Development of Ideas on the Structure of Metals before 1890* (Cambridge, Mass: MIT Press, 1988).

6. See, for example, *Materials Science and Engineering for the 1990s* (Washington D.C.: National Academy Press, 1989).

7. This data come from a graph in H.R Shubert's "Iron and Steel," in Singer, *History of Technology*, vol. 4, 106.

8. These numbers come from H. R. Shubert's "The Steel Industry," in Singer, *History of Technology*, vol. 5.

9. Robert Friedel, *A Material World: An Exhibition at the National Museum of American History* (Washington, D.C.: Smithsonian Institution, 1988), 49.

10. Gordon, *The Science of Structures*, 129.

11. S. B. Hamilton's "Building Materials and Techniques," in Singer, *A History of Technology*, vol. 5., 479. Steel had been used two years earlier in the upper floors of the first skyscraper ever built, also in Chicago. See Singer, *History of Technology*, vol. 5, 62.

12. On the history of ideas regarding the role of carbon in metals, see Smith, *History of Metallography*.

13. Siemens is quoted in W. H. G. Armytage, *A Social History of Engineering* (Boulder: Westview Press, 1976).

14. The details about Goodyear's quest are nicely laid out

by Stephen Fenichell in his *Plastic: The Makings of a Synthetic Century* (New York: HarperBusiness, 1996).

15. This account of the earliest synthetic plastics relies heavily on Robert Friedel's *Pioneer Plastic: The Making and Selling of Celluloid* (Madison: The University of Wisconsin Press, 1983).

16. Friedel points out that the camphor actually remains in the bulk of the material; see Friedel, *Pioneer Plastic*, 14–15.

17. Friedel, *A Material World*, 49–53. See also R. Chadwick, "New Extraction Processes for Metals," in Singer, *History of Technology*, vol. 5, 90–94.

18. Ibid.

19. Quoted from Matthew Josephson, *Edison*, (New York: John Wiley & Sons, 1959), 133–134.

20. David. A. Hounshell and John Kenly Smith, Jr., *Science and Corporate Strategy* (New York: Cambridge University Press, 1988), 2.

Chapter Three: The Secret Architecture of Stuff

The roots of the field of materials science and engineering are so many and varied that any account of them will by necessity reflect personal choices. Much of this chapter was pieced together from discussions and review articles appearing originally in venues such as *Scientific American* as well as many more technical books and articles. Still, there are several more general sources for finding parts of the story. Cyril Stanley Smith's *A History of Metallography: The Development of Ideas on the Structure of Metals before 1890* (Cambridge, Mass: MIT Press, 1988) is a superlative account of the beginnings of materials science before it even had a name. Parts of the narrative also can be picked out of Smith's previously mentioned *A Search for Structure* (Cambridge, 1981). An excellent, thorough, but very heavy-going source is entitled *Out of the Crystal Maze: Chapters from the History of Solid-State Physics* (New York: Oxford University Press, 1992), edited by Lillian Hoddeson,

Ernest Braun, Jurgen Teichmann, and Spencer Weart. A more concise and handier introduction can be found in Andrew Briggs, ed., *The Science of New Materials* (Cambridge, Mass.: Blackwell, 1992).

1. J. Randolph Kissell and Robert L. Ferry, *Aluminum Structures: A Guide to Their Specifications* and Design (New York: John Wiley & Sons, 1995).

2. Cochrane, Rexmond C., *Measures for Progress: A History of the National Bureau of Standards* (U.S. Department of Commerce: 1966).

3. Arden Bement, *Metallurgical Transactions A*, March 1987, 363–375.

4. Cyril Stanley Smith, personal communication.

5. Herman Mark, oral history interview archived at the Center for the History of Chemistry at the University of Pennsylvania.

6. Hermann Staudinger, *From Organic Chemistry to Macromolecules: A Scientific Autobiography Based on My Original Papers*, Herman F. Mark trans. (New York: Wiley-Interscience, 1976).

7. Herman Mark, oral history.

8. See, for example, P. W. Atkins, *Physical Chemistry*, 3d ed. (New York: 1986), 304.

9. This story has been told countless times in many different contexts. See, for example, S. Millman, ed., *A History of Engineering and Science in the Bell System: Physical Sciences (1925–1980)* (Short Hills, NJ: AT&T Bell Laboratories, 1983); and Hans Queisser, *The Conquest of the Microchip* (Cambridge, Mass.: Harvard University Press, 1988). Additional sources include Charles Weiner, "How the Transistor Emerged," IEEE *Spectrum*, January 1973, 24–33; and Mervin J. Kelly, "The First Five Years of the Transistor," *Bell Telephone Magazine*, Summer 1953, 72–86.

Part II—The Birth of a Superdiscipline

There are a number of first-person written accounts of the events leading up to and shortly following the birth of an academic discipline and field of research known as materials science and engineering. One of best began as a distinguished lecture given in 1986 by Arden I. Bement, Jr., at a gathering of materials researchers. Called "The Greening of Materials Science and Engineering" and appearing in the *Metallurgical Transactions A*, March 1987, 363–375. Two documents prepared by Roman J. Wasilewski were complementary to Bement's article. "Development of the Materials Research Laboratory Program in the U.S." was written shortly before its author passed away and was never published. The other was an "Outline of the MRL Program" that Wasilewski prepared at the end of 1981 for internal use within the National Science Foundation. These two documents were given to me by Lyle Schwartz, who heads the Materials Science and Engineering Laboratory at the National Institute of Standards and Technology. Another good source for this is Robert Sproull's "The Early History of the Materials Research Laboratories," in *Annual Reviews of Materials Science* (Palo Alto, CA: Annual Reviews Inc., 1987). *Advancing Materials Research* (Washington D.C.: National Academy Press, 1987) has several informative historical chapters by insiders.

1. William O. Baker, "Advances in Materials Research and Development," in (editors), *Advancing Materials Science*, eds. Petre A. Psaras and Dale Langford (Washington, D.C.: National Academy Press, 1987), 3–24. The advisory committee's paper is reprinted in full on page 24. Quote appears on page 4.

2. The two-age document that Fine gave the dean was entitled "Importance of Materials Science and Engineering."

3. A memo titled "Metallurgy Faculty Meeting" records the vote. Another memo, dated January 23, 1959, indicates that the relevant dean had approved the request to change the department's name.

4. Rustum Roy, in the twentieth anniversary coverage in the *MRS Bulletin* 18 (September 1993), 73–90.

5. *Materials Science and Engineering for the 1990s* (Washington, D.C.: National Academy Press, 1987).

Chapter Four: Materials Research Comes of Age

The portrait of a field that emerges in this chapter derives mostly from many personal interviews and visits with scientists and engineers and from reading their research and review articles. One official expression of how materials science has come of age since its inception is the book-length report *Materials Science and Engineering for the 1990s* (Washington D.C.: National Academy Press, 1989).

1. The examples of "ski-ramp" curves cited here are derived mostly from *Materials Science and Engineering for the 1990s.*

2. See, for example, K. Alex Muller and J. Georg Bednorz, "The Discovery of High-Temperature Superconductors," in *Nature*, 4 September 1987, 1133–1139.

3. Ivan Amato, "Turning Polymer Spaghetti into Lasagna," *Science*, 13 September 1991, 1212, reporting on presentation by Stupp at the Fourth Chemical Congress of North America in New York City.

4. Phaedron Avouris, personal communication.

5. L. Esaki and R. Tsu, "Superlattice and Negative Differential Conductivity in Semiconductors," *IBM Journal of Research and Development* (January 1970), 61–65.

6. See, for example, James L. Wilbur, Amit Kumar, Enoch Kim, and George Whitesides, "Microfabrication by Microcontact Printing of Self-Assembled Monolayers," *Advanced Materials*, 1994, 6: 600–604.

7. For a nontechnical review of Stoddart and his ilk, see Ivan Amato, "Designer Solids: Haute Couture in Chemistry," *Science*, 7 May 1993, 593–595.

8. For a review of biomimetic materials, see Ivan Amato,

"Heeding the Call of the Wild," in *Science*, 30 August 1991, 966–968.

9. Julian Vincent, a sociologist and materials scientist at Reading University in England, personal communication.

10. Quoted from *Impact of Supercomputing Capabilities on U.S. Materials Science and Technology* (National Material Advisory Board National Research Council, September 1988), 38.

11. Much of the discussion on theory and computational approaches is derived from the research that I did for the following article: Ivan Amato, "The Ascent of Odorless Chemistry," *Science*, 17 April 1992, 306–308. I spoke with a dozen or so practitioners of computational chemistry and materials science. Examples of primary sources include Henry F. Schaefer III, "Methylene: A Paradigm for Computational Quantum Chemistry," *Science*, 7 March 1986, 1100–1107; Marvin L. Cohen, "Predicting New Solids and Superconductors," *Science*, 31 October 1986, 549–553; D.D. Vedensky, S. Crampin, M. E. Eberhart, and J. M. Maclren, "Quantum Mechanics and Mechanical Properties: Towards Twenty-first Century Materials," *Contemporary Physics* 31 (1990): 73–97.

12. Marvin Cohen, personal communication.

13. Amy Y. Liu and Marvin L. Cohen, "Prediction of Low Compressibility Solids," in *Science*, 25 August 1989, 841–842.

14. Charles T. Casale and Bruce Gelin, *Growth and Opportunity in Computational Chemistry* (Boston: Aberdeen Group, 1992).

15. William Goddard III, personal communication.

16. Tommaso Toffoli and Norman Margolus, *Cellular Automata Machines* (Cambridge, Mass.: M.I.T. Press, 1987).

Chapter Five: The Materials Serengeti

Twice each year at semiannual gatherings of its members, the Materials Research Society in Pittsburgh publishes a phonebook-size tome filled with several thousand abstracts of the meeting's oral and poster presentations. These abstract books and meetings are a particularly relevant example of many similar books and meetings that reveal the same kind of Serengeti metaphor for the field. The three more focused subsections of the chapter are based almost entirely on personal interviews, visits, and the technical literature. What follows are several more general sources for each section.

Synthetic Diamond. See Robert Hazen's *The New Alchemists: Breaking Through the Barriers of High Pressure* (New York: Times Books, 1993). It is a nicely written account of the high-pressure, high-temperature route to synthetic diamond and touches lightly on the low-pressure route that is my focus in this chapter section. The Materials Research Society published a special issue of its journal *The Journal of Materials Research* 5 (November 1990) that included over three hundred pages of technical papers that collectively give a flavor of the field. A number of insiders have written informative review articles. One is by Robert C. DeVries: "Synthesis of Diamond Under Metastable Conditions," *Annual Review of Materials Science* (Palo Alto: Annual Reviews, 1987). John C. Angus also has presented excellent technical histories of the field. See for example, Angus and Cliff C. Hayman's article in *Science* (August 18, 1988, 913–921) entitled "Low-Pressure, Metastable Growth of Diamond and 'Diamondlike' Phases." Also see *Status and Applications of Diamond and Diamond-Like Materials: An Emerging Technology* (Washington D.C.: National Academy Press, 1990), another way to get a feel for the synthetic diamond arena.

Biomimetic Materials. I know of no general texts focusing on either biomimetic research or smart materials. A short review can be found in "Heeding the Call of the Wild" by Ivan Amato in *Science* (August 30, 1991, 966–968).

Smart Materials. To get your feet wet in smart materials, see Ivan Amato's chapter "Smart Materials" in *On the Cutting Edge of Technology* (Carmel, Indiana: SAMS, Publishing, 1993) and "Animating the Material World" by Ivan Amato in *Science*, (January 17, 1992, 284–286).

1. For this and plenty of other interesting historical revelations about the quest for artificial diamonds, see Robert Hazen, *The New Alchemists: Breaking Through the Barriers of High Pressure* (New York: Times Books, 1993).
2. Robert Wentorf, personal communications. Also, see Ivan Amato, "Diamond Fever," *Science News*, 4 August 1990, 72–74.
3. Rustum Roy, personal communication.
4. John Angus, "History and Current Status of Diamond Growth at Metastable Conditions," a talk presented at the 175th meeting of the Electrochemical Society in Los Angeles, California, on May 8, 1989.
5. John Angus, personal communication.
6. DeVries, personal communication.
7. Rustum Roy, personal communication.
8. John Angus, personal communication.
9. Rustum Roy, personal communication.
10. National Research Council, *Status and Applications of Diamond and Diamond-Like Materials: An Emerging Technology* (Washington D.C.: National Academy Press, 1990).
11. John Angus, personal communication.
12. H. Kagan, et al., "Diamas: A Compact Diamond-Based Detector for the SSC," unpublished document.
13. All quotes in this paragraph from DeVries, personal communication.
14. National Research Council, *Status and Applications*.
15. That idea formed the basis of Seitz's 1975 patent claim and showed up in Tom Clancy's techno-espionage thriller, *The Cardinal of the Kremlin*, in which Soviet military scientists,

with the help of brilliant kidnapped U.S. researchers, are on the verge of developing SDI-type lasers capable of destroying missiles and satellites from the ground.

16. Previous lab-made polymers such as celluloid were derived from existing natural polymers such as the cellulose of plants.

17. Julian Vincent, personal communication. Julian Vincent "The Design of Natural Materials and Structures," *The Journal of Intelligent Materials and Structures* 1 (January 1990): 141–145; Julian Vincent, "Relationship between Density and Stiffness of Apple Flesh," *Journal of Food and Agriculture* 47 (1989): 443–462.

18. See, for example, *Advanced Materials* 3 (1991): 449–452. This includes Rustum Roy's argument and a rebuttal by John Maddox, editor of *Nature*.

19. Steven Vogel, *Life's Devices* (Princeton: Princeton University Press, 1988), 315.

20. Eric Baer, personal communication. Also see, E. Baer, A. Hiltner, and H. D. Keith, "Hierarchical Structure in Polymeric Materials," *Science*, 27 February 1987, 1015–1022; Eric Baer, "Advanced Polymers," *Scientific American*, October 1986, 179–190. Eric Baer, James J. Cassidy, and Anne Hiltner, "Hierarchical Structure of Collagen Composite Systems: Lessons from Biology," *Pure & Applied Chemistry* 67 (1991): 961–973.

21. Paul Calvert, personal communication; Paul Calvert, "Biomimetic Materials," unpublished presentation at a conference on Intelligent Materials in Hawaii, 1990.

22. Mehmet Sarikaya, personal communication.

23. Greg Olson, personal communication.

24. Mehmet Sarikaya, Katie E. Gunnison, and Ilhan A. Aksay, preprint of "Mechanical Property-Microstructural Relationships in Abalone Shell."

25. James Economy, University of Illinois at Urbana, personal communication.

26. Stephen Mann has written many articles on the topic of biomineralization. See, for example, "Molecular Recognition in

Biomineralization," *Nature*, 10 March 1988, 119–124; "Molecular Tectonics in Biomineralization and Biomimetic Materials Chemistry," *Nature*, 7 October 1993, 499–505; "Crystal Engineering: The Natural Way," *New Scientist*, 3 October 1990, 42–47.

27. Lia Addadi and Stephen Weiner, "Control and Design Principles in Biological Mineralization," in *Angewandte Chemie*, vol. 31 (Weinheim international English edition, 1992), 153–169.

28. Robert R. Jones, "Dan W. Urry Chosen Scientist of the Year," in *Research and Development Magazine*, October 1988, 56–57.

29. Urry, personal communication.

30. Donald S. Stookey, *Journey to the Center of the Crystal Ball: An Autobiography* (Columbus: The American Ceramics Society, 1985), 13–15.

31. Donald S. Trotter, "Photochromic and Photosensitive Glass," *Scientific American*, April 1991, 124–129.

32. Richard Claus, personal communication.

33. See Craig Rogers, "Intelligent Material Systems—the Dawn of a New Materials Age," pamphlet (Blacksburg: Center for Intelligent Materials Systems and Structures at Virginia Polytechnic and State University: 1992). This is a good source to get a sense of Craig Roger's philosophical views on the field of smart materials and structures.

34. One way to a get sense of the variety of research projects that fall under the rubric of smart materials is to leaf through the proceedings of one of the field's meetings. Some of the examples cited are described in Gareth J. Knowles, ed., *Active Materials and Adaptive Structures: Proceedings of the ADPA/AIAA/ASME/SPIE Conference on Active Materials and Adaptive Structures, 4–8 November 1991, Alexandria, VA* (Philadelphia: Institute of Physics Publishing, 1992).

35. Craig Rogers, personal communication.

36. Ibid.

37. Submarine cloaking was a major topic at the ADPA/AIAA/ASME/SPIE Conference.

38. W. Kraschmer, L. D. Lamb, K. Fostiropoulos, and D. R. Huffman, "Solid C60: A New Form of Carbon," *Nature,* 27 September 1990, 354–358.

39. Richard Smalley, Internet posting of a presentation titled "From Balls to Tubes to Ropes: New Materials from Carbon" at a meeting of the South Texas Section of the American Institute of Chemical Engineers in Houston on January 4, 1996.

Chapter Six: The Rite of Atomic Masons

The bulk of this chapter is based on personal communications and visits with Federico Capasso and from a large stack of technical papers on the field of band-gap engineering. For an introduction to band-gap engineering, see "Band-Gap Engineering: From Physics and Materials to New Semiconductor Devices" by Federico Capasso in *Science,* 9 January 1987, 172–176; and Federico Capasso and Alfred Y. Cho, "Band-Gap Engineering of Semiconductor Heterostructures by Molecular Beam Epitaxy: Physics and Applications," *Surface Science* 299/300 (1994): 878–891.

1. See, for example, Charles Weiner, "How the Transistor Emerged," *IEEE Spectrum,* January 1973, 24–33; and Mervin J. Kelly, "The First Five Years of the Transistor," *Bell Telephone Magazine,* Summer 1953, 72–86.

2. This story has been told countless times in many different contexts. See, for example, S. Millman, ed., *A History of Engineering and Science in the Bell System: Physical Sciences (1925–1980)* (Short Hills, NJ: AT&T Bell Laboratories, 1983); and Hans Queisser, *The Conquest of the Microchip* (Cambridge, Mass.: Harvard University Press, 1988).

3. In addition to Capasso, I spoke with Jerome Faist, Barbara Sivco, and Alfred Cho.

4. A fascinating account of the early Corning work on optical fibers is in Ira Magaziner and Mark Patinkin, *The Silent War: Inside the Global Business Battles Shaping America's Future* (New York: Vintage Books, 1989), 264–299.

5. L. Esaki, and R. Tsu, "Superlattice and Negative Differential Conductivity in Semiconductors," *IBM Journal of Research and Development*, January 1970.

6. R. F. Kazarinov and R. A. Suris, *Sov. Phys. Semicond.* 5 (1971): 207.

7. Jerome Faist, Federico Capasso, Deborah L. Sivco, Carlo Sirtori, Albert L. Hutchinson, and Alfred Cho, "Quantum Cascade Laser," *Science*, 22 April 1994, 553–556.

Chapter Seven: Composing Steel

The bulk of this chapter was based on personal communications and meetings with Stephen Carr and Greg Olson and from a large portfolio of Olson's technical papers. To get a feel for what materials design means, see "New Steels by Design" by Greg B. Olson in *Journal of Materials Education* 11 (1989): 515–528. The fullest articulation of the approach appears in Olson, Carr, Jennings, and Jones, "Materials Design Initiative: An Integrated Research and Teaching Consortium for Systems Design and Materials," a proposal to the National Science Foundation's Division of Materials Research, 1994.

1. T. A. Stephenson, C. E. Campbell, and G. B. Olson, "Systems Design of Advanced Bearing Steel," in *Advanced Earth-to-Orbit Propulsion Technology 1992*, eds. R. J. Richmond and S. T. Wu, NASA Conference Publication 3174, vol. 2 (1992), 299–307.

2. This information comes from Olson's recollection of discussions with the friend from Rockwell International who told him about the problem with steel in the pump bearings.

3. Jon Van, "New Meaning to Phrase Tough as Steel,'" *Chicago Tribune*, 31 December 1989, 3.

Chapter Eight: The Stuff That Dreams Are Made Of

1. Mark Eberhart, personal communication.

INDEX